Sustaining Mobile Learning

AF695794

Mobile technologies are one of the fastest growing areas of technology in education. For learners, they offer an appealing opportunity to transcend teacher-defined knowledge and approaches by being able to access multiple, alternative sources of information anytime and anywhere. While the pace of engagement with and research into the educational applications of mobile technologies has picked up dramatically in the last decade, there is no consolidated view of how to sustain the practices or opportunities that are being explored.

Sustainability is a complex but crucial issue in mobile learning. Educational institutions are usually required to make substantial investments in mobile devices and associated technologies, and time and training to initiate mobile learning programs. Non-sustainable programs mean that investment will be wasted and further innovation will be threatened. The complexity of sustainable mobile learning programs is further exacerbated by the fast pace of change of digital technologies, where with every change, new possibilities are opened up and investments required. In addition, educators are still attempting to reconcile pedagogies of formal education with informal mobile learning. The book addresses these issues, with a particular focus on:

- exploring the challenges surrounding the sustainability of mobile learning in K-12 and higher education
- investigating the importance of sustaining mobile learning for diverse populations of students globally
- discussing theoretical models for the sustainability of mobile learning
- providing the reader with strategies for sustaining mobile learning.

Presenting new research alongside theoretical models and ideas for practice, the book will appeal to educators, researchers, academics, and postgraduate students in the fields of education and mobile learning, as well as those working in teacher education.

Wan Ng is Associate Professor in Technology Education and Science Education at the University of Technology, Sydney, Australia. She has researched mobile learning for over a decade and has published widely in this area, including an edited book on innovative mobile learning pedagogy.

Therese M. Cumming is Senior Lecturer in the School of Education at the University of New South Wales, Australia. As part of the School of Education's Special Education Research Group, Therese's research is focused on using iPads to support individuals with disabilities.

Routledge Research in Higher Education

Experiences of Immigrant Professors
Cross-Cultural Differences and Challenges, and Lessons for Success
Edited by Charles Hutchison

Integrative Learning
International research and practice
Edited by Daniel Blackshields, James Cronin, Bettie Higgs, Shane Kilcommins, Marian McCarthy and Anthony Ryan

Developing Creativities in Higher Music Education
International perspectives and practices
Edited by Pamela Burnard

Academic Governance
Disciplines and Policy
Jenny M. Lewis

Refocusing the Self in Higher Education
A Phenomenological Perspective
Glen L. Sherman

Activity Theory, Authentic Learning and Emerging Technologies
Towards a transformative higher education pedagogy
Edited by Vivienne Bozalek, Dick Ng'ambi, Denise Wood, Jan Herrington, Joanne Hardman and Alan Amory

Understanding HIV and STI Prevention for College Students
Edited by Leo Wilton, Robert T. Palmer, and Dina C. Maramba

From Vocational to Professional Education
Educating for social welfare
Jens-Christian Smeby and Molly Sutphen

Academic Building in Net-based Higher Education
Moving beyond learning
Edited by Trine Fossland, Helle Mathiasen and Mariann Solberg

University Access and Success
Capabilities, diversity and social justice
Merridy Wilson-Strydom

Reconsidering English Studies in Indian Higher Education
Suman Gupta, Richard Allen, Subarno Chattarji and Supriya Chaudhuri

Globally Networked Teaching in the Humanities
Theories and Practices
Edited by Alexandra Schultheis Moore and Sunka Simon

Higher Education Access and Choice for Latino Students
Critical Findings and Theoretical Perspectives
Edited by Patricia A. Pérez and Miguel Ceja

The Experiences of Black and Minority Ethnic Academics
A comparative study of the unequal academy
Kalwant Bhopal

Sustaining Mobile Learning
Theory, research and practice
Edited by Wan Ng and Therese M. Cumming

Narratives of Doctoral Studies in Science Education
Making the transition from educational practitioner to researcher
Edited by Shirley Simon, Christina Ottander and Ilka Parchmann

Sustaining Mobile Learning

Theory, research and practice

Edited by Wan Ng and
Therese M. Cumming

Routledge
Taylor & Francis Group
LONDON AND NEW YORK

First published 2016
by Routledge

2 Park Square, Milton Park, Abingdon, Oxfordshire OX14 4RN
711 Third Avenue, New York, NY 10017

Routledge is an imprint of the Taylor & Francis Group, an informa business

First issued in paperback 2017

Copyright © 2016 Wan Ng, Therese M. Cumming and the individual chapter authors for their contributions

The right of the editors to be identified as the authors of the editorial material, and of the authors for their individual chapters, has been asserted in accordance with sections 77 and 78 of the Copyright, Designs and Patents Act 1988.

All rights reserved. No part of this book may be reprinted or reproduced or utilised in any form or by any electronic, mechanical, or other means, now known or hereafter invented, including photocopying and recording, or in any information storage or retrieval system, without permission in writing from the publishers.

Notice:
Product or corporate names may be trademarks or registered trademarks, and are used only for identification and explanation without intent to infringe.

British Library Cataloguing in Publication Data
A catalogue record for this book is available from the British Library

Library of Congress Cataloging-in-Publication Data
Sustaining mobile learning : theory, research and practice / edited by Wan Ng and Therese M. Cumming.
pages cm
Includes bibliographical references.
1. Mobile communication systems in education. 2. Education–Effect of technological innovations on. I. Ng, Wan, 1955- II. Cumming, Therese M.
LB1044.84.S87 2016
371.33–dc23
2015004357

ISBN: 978-1-138-78738-4 (hbk)
ISBN: 978-0-8153-5749-0 (pbk)

Typeset in Sabon
by Cenveo Publisher Services

Contents

List of figures and tables

Figures

Tables

Contributors

Tim Best, senior researcher at Wexford Institute in Rolling Hills, California, is a life-long learner and educator. Best holds a PhD from the Ohio State University and has spent his career in a variety of educational settings, including public K–12 education, higher education, non-profit organisations, for-profit companies, and state and federal government. Recently, he served as project director for a federal grant for the Ohio Board of Regents to build technology capacity in Ohio institutions that prepare teachers. He served in a similar capacity for a federal grant at City University of New York. Best participated in a team of experienced architects and educators to write *Architecture for Achievement: Building Patterns for Small School Learning*, a practical path through the maze of decisions encountered in a school design or building conversion project. He designed and taught a professional practices survey course to all first-year MBA students at the Fisher College of Business at the Ohio State University. He served as small school coach for the University of Washington's Small Schools Project, working with the Seattle Public Schools. In 1994 Best established Ohio SchoolNet, a state agency charged with fitting 100,000 public school classrooms in the state for data, voice, and video connectivity. Best lives with his wife, Sharon, in Dublin, Ohio, and Rancho Palos Verdes, California.

Cheng-ping Chen is an Assistant Professor of the Department of Visual Design at Taipei College of Maritime Technology. She earned her PhD in the Department of Information and Learning Technology at the National University of Tainan. Her research emphases include instructional pervasive game, as well as issues related to communication media for teaching and learning purposes.

Therese M. Cumming, PhD is a Senior Lecturer in the School of Education at the University of New South Wales in Sydney, Australia. She has earned her PhD and MEd in Special Education (EBD) from the University of Nevada Las Vegas, and her BSED Special Education from Bloomsburg University of Pennsylvania. Her research interests include: educational technology, transitions in the lives of people with disabilities, emotional and behavioural disorders, and intellectual disabilities. Recent research projects in mobile technology include: using iPads to support students with language-based disabilities, using iPads to support adults with intellectual disabilities in being researchers, teachers' perspectives on student misbehaviour, and using technology to enhance social skills instruction for students with emotional and behavioural disabilities. She has authored over 30 publications, and presented her work at both national and international conferences. Prior to her university and research work, Dr Cumming worked for many years as a special educator and behaviour mentor.

Roselyn Dixon, PhD has been a special education teacher and a research academic involved with Special Education for over 25 years. She has published papers in the fields of social skills and behavioural interventions for people with a range of disabilities, including students with Oppositional Defiance Disorders and Autism Spectrum Disorders. More recently she has been actively involved in examining the relationship between digital technologies and pedagogy in special education and inclusive classrooms. She is a Senior Lecturer in Special Education at the School of Education, University of Wollongong, where she is the Postgraduate and Undergraduate Coordinator of Special Education.

Laurel Evelyn Dyson is a Senior Lecturer in Information Technology at the University of Technology Sydney, and President of anzMLearn, the Australian and New Zealand Mobile Learning Group. She is a qualified adult educator and has over 25 years of experience teaching in the university and adult education sector, where she seeks to create innovative ways of engaging students and improving learning outcomes using a highly student-centred and interactive approach. She first became interested in mobile learning in 2007 and since then has implemented mobile learning activities involving student-generated vodcasts and screencasts, mobile-supported fieldwork, interactive classroom activities using 3G phones, and improving classroom learning using iPads. Dr Dyson has over 75 publications which include research into mobile technologies in higher education, issues of ethics and sustainability in mobile learning, collaborative learning, and the adoption of mobile technologies by Indigenous people. She has provided leadership in mobile learning through her role as the founder of anzMLearn, which seeks to provide a forum for mobile learning researchers and practitioners throughout Australia and New Zealand. As president of anzMLearn, she conducts the annual Mobile Learning Research Workshop, and edits the group's journal *Transactions on Mobile Learning*. She is the recipient of five faculty, university and national teaching awards, including two Australian Office of Learning and Teaching Citations in 2012 and 2013.

Teresa Franklin, PhD is Professor of Instructional Technology in the Educational Studies Department at Ohio University. She has a passion for teaching and learning through the integration of technology into curriculum and instruction that spans 36 years. Research interests include: examination of online/e-learning environments, mobile technologies, virtual learning environments and technology integration in math and science. Dr Franklin has significant publications that integrate technology into science and math including *Teaching Science for All Children* (5 editions), *Virtual Games and Career Exploration*; *The Mobile School: Digital communities created by mobile learners*; and *The Changing Roles of Faculty and Students when Mobile Devices Enter the Higher Education Classroom*. She is the Director of the OU Global Services Program, which supports educational partnerships with universities worldwide. She was a recipient of the 2014 iNACOL Innovation Award for outstanding research in blended learning.

Tim Goodchild is currently a Senior Lecturer in the Faculty of Health and Science at University Campus Suffolk (UCS), UK. Tim began his career as a registered nurse working at Addenbrookes NHS Trust and Ipswich NHS Trust before moving into nurse and healthcare education. Tim is currently course leader for an MA in Healthcare Education (distance learning) and lectures on healthcare education, research methods, technology and nursing. In his time at UCS, Tim has led many projects including the exploration and use of virtual patients in distance learning, the realities of 'blended learning', and the use of social media, mobile learning, and reusable learning objects in healthcare education. Tim is currently studying for a PhD at Essex University titled 'The Fantasy of Technology Enhanced Learning', which utilises a discourse analysis and 'logics of critical explanation' approach to explore technology enhanced learning in nurse education, with a particular focus on the enduring promises of educational technology and the actual reality of technology in teaching and learning today. Tim has also published and spoken at several conferences in the area of healthcare education and technology enhanced learning.

Jyun-Fong Guo graduated with a Master's degree from Department of Information and Learning Technology at the National University of Tainan in Taiwan. His research focuses on mobile learning and pervasive game.

Hanne Wacher Kjærgaard is a senior lecturer at VIA University College, Denmark. She holds an MA in English and has trained English teachers for K–10 for many years, always working with ICT and its affordances and potential contributions to learning, publishing articles on these topics. Since 2004, Hanne has also been involved in VIA's research centre for ICT and learning, for the past 5 years as head of the centre. Hanne has been involved in and published on many projects centring on ICT and learning in schools, in higher education and in informal learning arenas, just as quality in e-learning and blended learning has been in focus.

Hanne has developed ICT-based books and feedback tools for schools and from 2015 will be working on a PhD on ICT-mediated feedback.

Lars Peter Bech Kjeldsen is Associate Dean of Research and higher degrees at VIA University College, Denmark, and holds an MSc in psychology and a PhD in ICT and organisational development. For a number of years he has taken an active interest in the development of technology in the field of education. His research work and publications have focused on the following themes: learning and supervision in virtual learning environments; knowledge sharing and communication using digital media. Lars plays an active role in research and development projects focusing on quality development and evaluation and has worked with ICT-based tests in teaching and ICT-based screening materials. In recent years, he has been involved in various research projects dealing with kinesthetic learning. This commitment has also resulted in a number of publications on this subject.

Agnes Kukulska-Hulme is Professor of Learning Technology and Communication in the Institute of Educational Technology at The Open University, UK. She has held academic positions in the fields of language teaching, computational linguistics and educational technology. Since 2001 her focus has been on mobile learning research and innovative practice. She is Past President of the International Association for Mobile Learning and serves on the editorial boards of several journals including the *International Journal of Mobile and Blended Learning*, and *ReCALL*, the research journal of the European Association for Computer-Assisted Language Learning. Agnes has experience of management and research on various UK and EU projects including MOTILL: Mobile Technologies in Lifelong Learning, the FP7-MASELTOV project on smart and personalised technologies for social inclusion, the British Council funded Mobile Pedagogy for English Language Teaching, and the SALSA project on immigrant language learning in the next generation of smart cities. Agnes also contributes to The Open University's Master's program in Online and Distance Education.

Chee-Kit Looi is Professor of Education in the National Institute of Education, Nanyang Technological University, Singapore. He was the Founding Head of the Learning Sciences Lab, Singapore from 2004 to 2008, the first research centre devoted to the study of the sciences of learning in the Asia-Pacific region. His research interests include mobile and ubiquitous technologies, and computer-supported collaborative learning. He serves on the editorial boards of the *IEEE Transactions in Learning Technologies*, *International Journal of AI in Education*, *International Journal of CSCL*, and *Journal of the Learning Sciences*.

Wan Ng, PhD is Associate Professor in Technology Education and Science Education at the University of Technology, Sydney, Australia. Her technology education research interest lies largely in innovative pedagogies enabled by digital technologies that have authentic and sustainable impact on students' learning. She is also interested in how young people interact with technologies to learn. Since 2002, the start of her academic career in education, Wan has attracted numerous federal, state and university grants and consultancy work for innovative projects in the areas of technology education and science education. One of her current projects is the multi-institutional Australian Government-funded $1.64 million *Smart Science Initiative* project that embraces a gamified, adaptive learning platform to engage years 9 and 10 students with science learning, and where the use of tablet technology in schools is a common feature of the project. Wan has produced a sole-authored book titled *Empowering Scientific Literacy through Digital Literacy and Multiliteracies* and a sole-edited book titled *Mobile Technologies and Handheld Devices for Ubiquitous Learning: Research and Pedagogy.* She has written more than 60 refereed publications for an international audience and is currently under contract to author a book on professional learning in new digital technologies for educators.

Howard Nicholas is a Senior Lecturer in Language Education at La Trobe University (Australia) and in 2013–4 was Visiting Professor of English Language Education at Paderborn University (Germany). His research interests include second language acquisition and plurilingualism, language policy and multilingual education. His work with Wan Ng has explored diverse aspects of learning with mobile devices in schools and higher education including studies of how learners' use of mobile devices can capture their thinking and the challenges for schools and universities in attempting to introduce and sustain such programs. His central interest is how learners gain control over their own learning by using mobile devices – a central theme in his work with Caroline Walta. His work in mobile learning contributed to his recently published book (with Donna Starks) *Language Education and Applied Linguistics: Bridging the Two Fields* (Routledge, 2014), which elaborated 'multiplicity' as a framework for understanding and working with plurilingual and multimodal communication.

Annette Rahn is a lecturer at VIA University College, Denmark, and holds an MSc in Nursing. Since 2012, she has been a lecturer at the VIA School of Nursing at Horsens, Denmark, where she teaches human anatomy and physiology for nursing students. Annette has been involved in projects about Augmented Reality (AR) and lung anatomy in 2012 and 2013. In 2014 she co-authored the article 'Augmented Reality as a visualizing facilitator in nursing education'. Her focus has been the learning potential in using AR technology to facilitate the acquisition of complex knowledge in natural science subjects through visualisation and interaction. In 2015 she will continue with a project about AR technology and lung anatomy as an innovative method for the acquisition of human anatomy and physiology.

Andy Ramsden is currently the e-Learning Development Manager at UCS. The responsibilities within this role include feeding into and deploying institution-wide technology enhanced learning strategies. This involves operational management over staff and student development programmes, e-learning technologies and evaluating impact. Within this capacity he manages the e-learning team (http://ucselevate.blogspot.co.uk). Previous to this post he was the Head of e-Learning at the University of Bath and the VLE Operational Manager and Learning Technology Advisor at the University of Bristol. His current teaching portfolio at UCS includes lecturing on the Postgraduate Certificate in Academic Practice, and on the MA in Learning and Teaching. Mobile learning, including the use of PDAs in Initial Teacher Training, and student attitudes to the use of handheld devices as learning technologies, has been a long-standing research interest, and he has been writing in this area for a number of years. For more details see https://ucs.academia.edu/ARamsden. Andy has had a wider role within Technology Enhanced Learning within the UK, as a long standing steering group member of the Heads of e-Learning Forum (http://helfuk.blogspot.co.uk/). To find out more, visit is professional blog (http://andyramsden.wordpress.com) or follow him on Twitter (andyramsden).

Mike Sharples is Professor of Educational Technology in the Institute of Educational Technology at The Open University, UK. He also has a post as Academic Lead for the FutureLearn company. His research involves human-centred design of new technologies and environments for learning. He inaugurated the mLearn conference series and was Founding President of the International Association for Mobile Learning. He is Associate Editor in Chief of *IEEE Transactions on Learning Technologies*. He is author of over 300 papers in the areas of educational technology, science education, human-centred design of personal technologies, artificial intelligence and cognitive science.

Ju-Ling Shih is a Professor in the Department of Information and Learning Technology at the National University of Tainan, Taiwan. She earned her EdD in Communication and Education from Teachers College, Columbia University. She has two Master's degrees: EdM in Communication and Education from Teachers College, Columbia University; and MS in Broadcasting Production from Boston University. Her research interests include instructional design and qualitative research in digital learning, digital games, mobile learning, and technology-mediated education in various levels and fields. She developed a cross-platform 3D role-play digital learning game, "Taiwan Epic Game", that integrates mobile-learning and digital games for the exploration of Taiwan history and culture. Her publications can be seen in academic journals such as *Educational Technology and Society*, *Computers & Education*, and *British Journal of Educational Technology* among others. She received the 2012 Ta-You Wu Memorial Award, and 2011–2013 Excellent Young Scholars Project from Ministry of Science and Technology, Taiwan.

Iva Strnadová is Associate Professor of Special Education in the School of Education, Faculty of Arts and Social Sciences, University of New South Wales. She is also an honorary staff member at the University of Sydney, Faculty of Education and Social Work, Australia. Her research interests include life-span transitions experienced by people with disabilities, well-being of people with developmental disabilities (intellectual disabilities and autism) and their families, women with intellectual disabilities, and mobile learning for people with developmental disabilities. She has authored 95 publications across all of the publication categories, and has been awarded 19 competitive grants in Australia and internationally.

Daner Sun is a research fellow in the National Institute of Education, Nanyang Technological University, Singapore. From April 2015, she will be an Assistant Professor in the Hong Kong Institute of Education, China. Her research interests include ICT-supported science learning, science curriculum design and development and science teacher education. She serves as a reviewer of journals in science education and educational technology.

Irina Verenikina, Associate Professor, is an educational psychologist in the School of Education, Faculty of Social Sciences, at the University of Wollongong. She is a Full Member of the Australian Psychological Society. She holds a PhD in Developmental Psychology from the Russian Academy of Education and an MSc (Honours) in Educational and Developmental Psychology from the Faculty of Psychology, Moscow State University, Russia. Before moving to Australia she held a senior research fellow position in the Russian Academy of Sciences, International Laboratory of Communication and New Technologies. Her research interests relate to the application of Sociocultural Psychology and Activity Theory to the study of the effective use of digital technologies in teaching and learning in various educational contexts, such as early childhood education, special education, literacy and music education. She is the author of over 100 publications. In 2014 A/Prof Verenikina chaired the Triennial International Congress of the International Society for Cultural Research and Activity Theory (ISCAR) which was held in Sydney, Australia. From 2003–2008 she served as an elected Regional Coordinator for Australia and New Zealand at the International Executive Committee of ISCAR. In this capacity she led the 1st and 2nd Regional International Conferences in Australia (in 2004 and 2007).

Caroline Walta is a senior lecturer at La Trobe University, Shepparton campus. She coordinates a postgraduate program for teacher accreditation which is accessed in blended learning mode involving online learning, intensive on-campus workshops and virtual tutorials. Her research interests involve technology and learning and the management of practicum experience, in rural and remote locations and internationally. Recent publications have involved a year-long research project which measured the impact of the use of the iPod Touch as a device for engagement with learning in the teacher education program and an overview of the theory and practice of program design to support online learning programs.

Scott Walthour, EdTech Mentoring Group. Scott has a diverse educational technology background that includes classroom teaching, curriculum coordination, grant writing, project management and technology leadership in K–12 and higher education. He has directed technology efforts at two private universities and held leadership positions in Ohio's K–12 educational technology agencies, including serving as Chief Learning Officer of eTech Ohio, Director of Client Affairs at Ohio SchoolNet and program manager for Ohio's Technology Literacy Challenge Fund. Prior to these experiences he led two successful county-wide technology projects and managed a ten-county mentorship program for first-year teachers while serving as Putnam County's (Ohio) Secondary Curriculum Coordinator. For 11 years prior to this experience he was a secondary social studies teacher at Leipsic Public Schools. Currently, Scott runs his own educational consulting service, working with public and private sector clients. This work includes program managing a statewide roster verification initiative associated with evaluating student growth, serving on a private telecommunications company board and the institutional review board of a non-profit educational consulting company. He resides with his wife of 34 years in Ottawa, Ohio.

Lung-Hsiang Wong is a Senior Research Scientist in the National Institute of Education, Nanyang Technological University, Singapore. His research interests are mobile and seamless learning, computer-assisted language learning, computer-supported collaborative learning and teachers' professional development. He is an Associate Editor of *IEEE Transactions on Learning Technologies*. He was the lead editor of the scholarly book *Seamless Learning in the Age of Mobile Connectivity*, (Springer, 2015), and the first author of the Chinese scholarly book *Move, Language Learning! – Exploring Seamless and Mobile Language Learning* (Nanjing University Press, 2011).

Longkai Wu is currently a Research Scientist and faculty member at the Office of Education Research, National Institute of Education, Singapore. He holds a PhD in the area of Learning Sciences and Technologies from Nanyang Technological University (NTU), Singapore. His current research is on the development of learning technologies that can be translated and scaled into classroom practices.

Foreword

Many years ago, in the heyday of innovation, I used to say that funders should stop supporting good projects in the hope that they would become sustainable and instead support sustainable projects in the hope that they would become good. Unfortunately, but understandably, the nascent mobile learning research community often bought into the lexicon and methods of innovation as understood by the e-learning community and its funders, looking for critical mass, early adopters and diffusion of the innovations. Apart from the general failings of innovation, the innovative projects of mobile learning often supplied learners with mobile devices, in the interests of methodological and operational simplicity. This was demonstrably not sustainable, however positive and convincing the evidence and the findings.

Times have changed, the devices that were originally expensive, scarce, difficult and fragile have become easy, cheap, universal and robust but the challenge of sustainability seems only to have transmuted rather than disappeared. The transition from supplied devices to BYOD is seductive but not widely established. The current book represents the first concerted attack on the challenge to appear in the literature and tackles that challenge from a rich variety of angles. It deserves to be widely read, outside the research communities, because learning with mobiles is no longer an optional successor to learning with computers but the authentic mode of learning for societies characterised by mobility and connection.

This suggests that learning with mobiles ought, almost by logical necessity, to be sustainable because it is the learning of a mobile and connected society. Perhaps there is an evitable lag before the professions and institutions of established education systems catch up with this realisation and before the mobile learning research community shifts its focus from evidence and projects to policy and deployment.

This may, however, be something of an over-simplification; there will always be a place for innovation and experimentation, though less and less technological and more and more pedagogic, and researchers already recognise the need to engage with policy, policy makers and the wider public as well as with refereed journals, professional bodies and learned societies.

Many years ago, I also used to show a slide of the biblical parable of the seed falling on stony ground. We had been preoccupied with the seed and not with the soil, with the specifics of technological and pedagogic innovation and not with the less tractable details of the host institution, community or organisation, and this, in my over-simplified rhetoric, explained our problems with sustainability. Times have changed and now the social and organisational contexts of mobile learning form part of a more balanced account of mobile learning. The current book addresses sustainability in far richer and more varied senses than perhaps we and our funders saw it all those years ago.

There are an ever-increasing number of books on mobile learning coming out, some still generalist, other specialist, exploring for example maths education or contextual education. They will all, however, depend on a better understanding and delivery of sustainability if they are to have any practical significance.

This is one sign of increasing maturity, breadth and sophistication in the mobile learning community but at the same time we are seeing a change in scale, emphasis and distribution. Whilst the global economic downturn may have had a negative impact on research funding, shifting some of the emphasis away from speculative research, it has also made sustainability more urgent and important.

It could be argued that the arrival of the iPhone in 2007 catalysed and epitomised the US interest in the mobile (or vice versa perhaps) and led to an awareness and an interest amongst US agencies, such as USAID and the Peace Corp, amongst US foundations, such as Ganz Cooney, Macarthur and Hewlett Packard, and amongst international organisations staffed mainly by US nationals, such as the World Bank, World Economic Forum, UNESCO, UNHCR, the International Telecommunications Union and the International Labour Organisation, to deliver their various humanitarian and educational missions in the global South. This was accompanied by a greater awareness amongst corporates, for example, Nokia, Pearson and the MNOs represented by the GSMA, of the commercial potential educational offers to the global South in their portfolios, the GSMA talking of the 'next billion subscribers' and of 'A Platform for Educational Opportunities at the Base of the Pyramid' in 2010. This period also saw the emergence of the apps economy, and the growth of user-generated learning, not only apps across every major platform but e-books, blogs, web-sites and communities, often originating outside the domains of commercial developers, professional educators and professional researchers. This marked a shift in the balance of mobile learning activities as well as an enormous increase in scale. In the broadest terms, these changes demonstrated increasing diversity in expressions of sustainability and in underlying business models for mobile learning. There was also increased pressure from amongst the newly interested policy, commercial and donor communities in scale and sustainability which was not wholly benign, together and separately representing a threat or a challenge to the earlier pedagogic richness and perhaps the primacy of educational concerns.

The shift in the centre of gravity towards the US may have produced a greater interest in games, drill and training at the expense of theoretically informed informal learning. The imperative to scale may have shifted the emphasis towards models of learning based around content and its delivery and away from models based on discussion, community and connection. The imperative to scale may have favoured formal learning over informal learning and education for the mainstream over education for the marginal.

The emergent players may have reinforced the existing monopoly of US English as the preferred medium and at the same time squeezed out researchers as mobile technologies became accessible, easy and universal. Research funding, the nature of the funding ecosystem, the dominant research questions and the early technical challenges may have diverted attention away from sustainability and the occasional donations from corporates (CSR, often short-term, often in the form of air-time or handsets) skewed perceptions about the economics of sustainability.

This rather personal, informal and impressionistic account illustrates the need for continued rigorous exploration of the sustainability of mobile learning from the research community, of the kind encapsulated in the current excellent volume.

Professor John Traxler
University of Wolverhampton

Preface

While the use of digital technologies in education started about half a century ago, the pace of change in the evolution of educational technologies has never been greater than the last 15 years. One of the fastest areas of growth is mobile technologies. According to the 2013 International Telecommunication Union report,[1] the mobile phone subscription figure was around 6.8 billion in 2012. As smartphones become more affordable and as 3G and 4G networks become more advanced, the switch from mobile phones to smartphones has also increased rapidly. eMarkerter[2] estimated that there are 1.75 billion users (or 38.5% of mobile phone users) using smartphones in 2014 while a 2013 Ericsson mobility report[3] estimated that by 2019, smartphone subscriptions will reach 5.6 billion.

The ownership of smart mobile devices (smartphones and computer tablets) by adolescents and young adults is also on the rise, with reports indicating that the ownership is 66% or higher in countries such as Australia, Hong Kong, Singapore, South Korea, Taiwan, United Kingdom and United States (see Chapter 5). As the affordability and capabilities of these mobile devices increase, they are becoming more attractive as devices for learning in formal education. This is evident in the growing interest, globally, in the concept of the bring-your-own-device (BYOD) model shown by schools and higher education institutions. Depending on the model of BYOD adopted, we see a potential shift in the mobile learning paradigm from single platform-based mobile technology use to multi-platform devices with non-uniform capabilities. There are challenges associated with this shift and with mobile learning in general, as discussed by Wan Ng and Howard Nicholas in Chapter 1 of the book.

For learners, however, smart mobile devices offer an appealing opportunity to transcend teacher-defined knowledge or approaches by being able to access multiple, alternative sources of information anytime and anywhere, that is, whenever the need arises. Mobile devices in this context are handheld, multipurpose and easily portable devices with Internet access. They include smartphones, tablets, PDAs, and iPods. While the pace of engagement with and research into the educational applications of mobile technologies have picked up dramatically in the last decade, there is no consolidated view of how to sustain the practices or opportunities that are being explored. Sustainability of mobile learning is a crucial issue since educational institutions are usually required to make substantial investments in mobile devices and associated technologies (e.g. infrastructure and apps), educators' time in re-designing curriculum to incorporate mobile learning, technical support, and professional development to initiate mobile learning programs. Non-sustainable programs mean that the investment will be wasted and further innovation threatened. The challenges of sustaining mobile learning programs are further exacerbated by the fast pace of change of digital technologies, where with every change, new possibilities are opened up and investments required. In addition, educators are still attempting to reconcile formal educational practices with informal mobile learning. The book aims to address these issues by: (i) exploring the challenges surrounding the sustainability of mobile learning in K-12 and higher education; (ii) conceptualising mobile learning and theoretical models for its sustainability; and (iii) providing readers (educators, researchers, instructors, principals, policy makers, IT personnel and undergraduate and higher degree students) with a rich set of arguments for sustaining mobile learning across various contexts of education that includes both theoretical perspectives and practical strategies for its realisation.

Sustainability of mobile learning is a complex issue. In Chapter 2, Tim Best, Scott Walthour and Teresa Franklin allude to one aspect of this complexity by referring to the irony in discussing sustaining mobile learning since mobile learning itself is a moving target. The researchers in the book approach this predicament in different ways by offering diverse understandings of what sustaining mobile learning means in their various educational contexts. The authors' views of sustainability in the book focus on:

- *Institutional organisation and structure.* In Chapter 1, Wan Ng and Howard Nicholas discuss a person-centred model for sustaining mobile learning in educational institutions. In the model, the authors consider human factors directly and discuss how stakeholders interacting with each other and with the technology are important perspectives to consider in theorising sustainable mobile learning programs. They show how the elements of this model can be used to identify five kinds of challenges (economic, political, technical, social and pedagogical) to the sustainability of learning with mobile devices. In Chapter 2, Tim Best, Teresa Franklin and Scott Walthour propose a model with structures that provoke exploration and invention rather than prescribe a discrete, bounded process, where they emphasise learning as a shift from knowing about something to applying new knowledge in context to create something. They called for a reinvention that disrupts the conditions that supports the status quo including the role of the teacher; the linear, time-based class; the heavy dependence on lecture and text; the emphasis on discrete, bounded factoids; the focus on prescribed content and individual learning; and the need for new assessments that demonstrate and apply conceptual mastery.

- *The life of the learner and lifelong learning.* In Chapter 3, Agnes Kukulska-Hulme and Mike Sharples show how learning can be sustained through learning journeys, since these give an extended sense of continuity and progress. Mobile and ubiquitous technologies can implement a seamless continuity of learning within and across contexts, offering additional ways of combining learning in the classroom, at home and outdoors. In Chapter 5, Wan Ng puts forward an argument that the sustaining aspect of using mobile devices for both learning and personal growth is an individual's mobile digital literacy. Mobile digital literacy integrates an individual's cognitive, technical and social-emotional ability to use mobile technologies for safe and effective learning across the individual's lifespan. It empowers the individual to adapt to new technologies easily in an era where technology is evolving at a rapid pace.
- *Professional development of educators.* In Chapter 4, Chee-Kit Looi, Daner Sun, Longkai Wu and Lung-Hsiang Wong discuss a conceptual model of Teacher Professional Development (TPD) that is critical in the lead-up to school-wide implementation of sustainable mobile learning. TPD was developed concurrently and iteratively with the development of innovative curricula over five years and encompass principles of effective professional development for teachers.
- *Innovations in mobile learning across different educational contexts.*

    - *Healthcare/nursing education*: in Chapter 6, Tim Goodchild and Andy Ramsden reflect critically on three years of use of educational podcasts in the School of Nursing at University Campus Suffolk and argue that knowledge and support of the diversity of staff experiences, skills, motivations and capabilities are issues critical to developing and sustaining the innovative and acceptable use of (mobile) technology in teaching and learning in healthcare education. In Chapter 10, Hanne Wacher Kjærgaard, Lars Peter Bech Kjeldsen and Annette Rahn investigate the potential of an iPad-facilitated application of augmented reality in the teaching of highly complex anatomical and physiological subject matter in nursing education. They present a case study and discuss issues and factors that sustain mobile-facilitated augmented reality application in nursing education.
    - *Special needs education*: in Chapter 7, Therese M. Cumming, Iva Strnadová, Roselyn Dixon and Irina Verenikina detail the potential that mobile learning under the Universal Design for Learning framework has for students in inclusive environments. They provide examples from the literature and their own research to illustrate effective implementation processes that promotes sustainability in mobile learning with special needs students.
    - *Teacher education*: in Chapter 8, Caroline Walta and Howard Nicholas present data from a pre-service teacher education program

that has learning with mobile devices as its core feature and has so far been sustained for seven years. Using the Community of Inquiry model, seven principles that describe how the program evolved over the seven years while sustaining its core focus are identified.

- *Primary education through gaming*: in Chapter 9, Ju-Ling Shih, Jyun-Fong Guo and Cheng-ping Chen argue that mobile learning is more sustainable through the use of pervasive gaming strategies. They detail a study that made use of the Tainan historical monument in Taiwan, Provintia, as the activity of the game, *Facebook* as the platform for group cooperation and smartphones with augmented reality capability for scene detection to enhance and sustain learning effectiveness and motivation.
- *Higher education*: In Chapter 11, Laurel Evelyn Dyson presents two innovative approaches to sustaining mobile learning in higher education. These are the BYOD and student-generated multimedia content approaches, both of which are widely accepted as learner-centred pedagogies. By using two case studies that have achieved sustainability for more than five years, the author discusses the sustainability of the approaches economically, technologically and pedagogically.

The chapters inform educational leaders and policy makers of issues associated with the implementation and sustainability of mobile learning programs, and the means of overcoming them. For educators, the chapters offer a broad, contextual perspective of theories and pedagogical strategies that they are able to adopt to ensure the sustainability of their practices in mobile learning.

We would like to thank Professor John Traxler for introducing the notion of *sustainability* in the Foreword and discussing the need for the research community to have continued rigorous exploration of this notion in terms of mobile learning. We also thank the authors for their varied perspectives on sustaining mobile learning in their chapters. Finally, we would like to thank the School of Education, University of New South Wales for supporting the research workshop on *Sustaining Mobile Learning* that provided the impetus for this book.

Wan Ng
Therese M. Cumming

Notes

1 http://www.itu.int/en/ITU-D/Statistics/Documents/facts/ICTFactsFigures2013-e.pdf
2 http://www.emarketer.com/Article/Smartphone-Users-Worldwide-Will-Total-175-Billion-2014/1010536
3 http://www.ericsson.com/res/docs/2013/ericsson-mobility-report-november-2013.pdf

1 Sustaining innovation in learning with mobile devices

Key challenges

Wan Ng and Howard Nicholas

Introduction

The penetration of mobile technology into our everyday lives is evident everywhere we look. According to a World Bank report, there were more than 6 billion mobile phone subscriptions worldwide in 2011 (World Bank, 2012). The International Telecommunication Union reported the subscription figure to be around 6.8 billion for 2012 (ITU, 2013). Kelly and Minges (2012, p. 8) in the World Bank report declared that this global pace of growth in mobile phone uptake is "unmatched in the history of technology", and also noted that the increase in adoption of mobile devices in developing countries is greater than in the developed world. For users in developing countries, there is now almost limitless accessibility to mobile applications (apps), at least in those places where apps have also been locally developed to address challenges such as affordability and literacy. Kelly and Minges (2012) further added that we are now only seeing the beginning of the growth curve in mobile technology and that with mobile devices becoming more powerful and affordable, the 'apps' economy will continue to grow, with a prediction of an 18-fold increase in mobile data traffic between 2011 and 2016 (Cisco, 2012).

While mobile devices are creating unprecedented opportunities for employment, education, and entertainment in developing countries, they are also making a great impact in developed countries, in particular through smartphones and tablets. In the 2010s, our lives have become comprehensively integrated into the digital world. The ITU (ITU, 2014) has predicted that in developed countries mobile-broadband penetration, that is, wireless Internet access through a mobile device, will reach 84% (of their populations) by the end of 2014. Patrik Cerwall (2012), Head of Strategic Marketing and Intelligence at Ericsson, stated that as the connected and networked society is taking shape, mobility is becoming an integral part of our everyday lives and future. The mobile phone has evolved from a single channel for voice with 2G (second generation wireless telephone technology) capability for text messages, picture and multimedia (MMS) messages, to the smartphone with 3G/4G wireless mobile Internet services and additional capabilities for instant text messaging, voice messaging and video calling features, streaming, global positioning systems (GPS), geotagging and high end applications. Table 1.1 shows Minges' (2012, p. 15) summary of the evolution of the capabilities of mobile devices from the basic mobile phone to the smartphone and computer tablet. The list of capabilities is not exhaustive, and not all devices have all features. However, the increased capabilities with more processing power of mobile devices has seen a dramatic rise in sales and ownership of smartphones and tablets over the last few years (BBC, 2008; Cerwall, 2012; Griffith, 2013a; Lomas, 2013; Nielsen, 2010; Whitney, 2009). BBC online (2008) reports that mobile net users are younger and search for different things on the Internet, and Nielsen (2010, p. 2) states that "Young people around the world are more immersed in mobile technology than any previous generation". Brown and Diaz (2010) highlighted the computing industry's focus on mobile technology, citing Google's CEO Eric Schmidt's assertion of "mobile first" in the company's policy. Innovations will continue to expand with changes in mobile devices, for example, attempting to capitalise on the best of both worlds from a phone and tablet wrapped into one to produce

Table 1.1 Mobile devices and their capabilities (adapted from Minges, 2012, p. 15)

Device	*Capabilities*	*Device*	*Capabilities*
Basic mobile phone	Network services, including: - Voice telephony and voice mail - SMS (short message service) - USSD (unstructured supplementary service data) - SMS-based services such as mobile money - USSD services such as instant messaging	**Smartphone**	As *Feature phone* plus: - Video camera - Web browser - GPS (global positioning system) - 3G+ internet access - Mobile operating "platforms" such as iOS, Android, Blackberry - Ability to download and manage applications - VoIP (Voice over Internet Protocol) - Mobile TV (if available) - Removable memory card
Feature phone	As *basic mobile phone* plus: - Multimedia Messaging Service (MMS) - Still picture camera - MP3 music player - 2.5G data access	**Tablet**	As *Smartphone* plus: - Front and rear-facing video cameras (for video calls) - Larger screen and memory capability - Faster processor, enabling video playback - Touchscreen with virtual keyboard - USB (universal serial bus) port

the 6-inch *phablet* (Griffith, 2013b). The phablet is a smartphone with the advantage of a larger screen that is intermediate in size between a typical smartphone and a tablet computer. Another innovation reported recently is Google's development of a tablet application to capture precise three-dimensional images of objects (Mukherjee, 2014).

In education, studies and reports of adolescents and young adults' ownership of smart devices are consistent with the reports of the rapid rise in sales of smartphones and tablets. For example, a recent study of 1,129 Australian adolescent students (ages 12–15) shows that 97% of the responding students owned a mobile phone at the end of 2013, with two-thirds (66%) owning a personal smartphone and 71% owning or having access to computer tablets at home or as part of their school programs (Ng & Nicholas, manuscript in preparation). The data is consistent with the report from UK's telecom regulator (Ofcom, 2013) on UK adolescents' proportion of ownership of mobile phones and tablets. The Ofcom report indicated that the smartphone ownership for 12–15 year olds was 62% and that there had been a tripling of tablet computers in the homes of 5–15 year olds, a sharp increase from their 2012 survey. Pew Internet Research in the US (2013) found a similar proportion of 66% smartphone ownership by young adults between 18–29 years of age. It is anticipated that almost all adolescents and young adults will own a smart device in the next few years, with the ownership starting at a much younger age. In a recent study *Zero to eight: Children's media use in America 2013* (Rideout, 2013) conducted by Common Sense Media, it was found that there had been a fivefold increase in tablet ownership in families with children aged 8 and under, from 8% in 2011 to 40%. The study surveyed 1,463 parents of children aged 8 and under and found that in 2013, 75% of these children had access to a smart device such as a smartphone and/or tablet. A similar number of these young children was found to use a mobile device for media activity such as watching videos or using apps. The research also found that the percentage of children under 2 years of age using mobile devices for media consumption had increased from 10% in 2011 to 38% in 2013. Children from lower-income families also had much better access to mobile devices, up from 22% in 2011 to 65% in 2013. The increase in access to digital technology, in particular mobile devices for young children means that by the time they begin schooling, many children would have explored technology and developed technology skills through manipulating games and processing information while viewing videos or interacting with educational apps, and socialising by chatting with grandparents and relatives through *Skype*, *Facetime* or equivalents.

The increase in accessibility to and convenience in using smart mobile devices for services enabled by the capabilities listed in Table 1 make them increasingly important in our lives. As Kelly and Minges state:

> Given technological convergence, mobile handsets can now function as a wallet, camera, television, alarm clock, calculator, address book,

> calendar, newspaper, gyroscope, and navigational device combined. The latest smartphones are not just invading the computer space, they are reinventing it by offering so much more in both voice and nonvoice services.
>
> (Kelly & Minges, 2012, p. 4)

Educators and policy makers are acknowledging that as part of this integration of smart mobile devices in people's lives, schools and higher education should capitalise on their capabilities and students' familiarity with the technology to innovate learning with these devices (New South Wales Department of Education and Communities (NSWDEC), n.d.). One challenge is designing a response to the rapidly changing technical landscape that we have described in this section. One response can be seen in the bring-your-own-device (BYOD) concept, which is being explored globally in education institutions. In the BYOD approach we are beginning to see a shift in the mobile learning paradigm from single platform-based mobile technology use to multi-platform devices with non-uniform capabilities being adopted by schools and higher education institutions. The development of such a diverse strategy has the capacity to make innovation more financially sustainable (schools pay less for the technology itself), but in other aspects more difficult to sustain (more complex support and more challenges for teachers). We explore this issue below as an example of how sustainability can be operationalised and possibly also achieved, but first we set the scene with a discussion of learning with mobile devices and the challenges in sustaining programs where mobile devices are used.

Mobile learning

While the literature on mobile learning has increased significantly in the last ten years to reflect the interest in integrating mobile learning into education (see literature reviews of Hwang & Tsai, 2010, 2011; Ng & Nicholas, 2013; Petrova & Li, 2009; Wu et al., 2012), there is no unified definition of the concept. Conceptualising mobile learning has varied. Some researchers have focused on access, mobility and support for learning in multiple locations without physical network connections (e.g. Georgiev, Georgieva, & Smrikarov, 2004; Parsons & Ryu, 2006). Others have used contexts, space and time to define mobile learning (Pachler, 2009; Sharples, Taylor & Vavoula, 2007), for example Pachler (2009) asserted that mobile devices enable users to "re-interpret their everyday life contexts as potential resources for learning" (p. 5). Other researchers have emphasised the capacity to bridge formal and informal contexts and the related potential ubiquity of mobile learning (e.g. Clough et al., 2009; Evans, 2008; Goodchild & Chenery-Morris, 2011; Kukulska-Hulme et al., 2009; Looi et al., 2010). Traxler (2007) asserted that it is these flexible attributes of mobile learning that make it different from traditional learning. He argued that the empowerment of personal learning experiences and the enabling convenience for learning make defining mobile learning difficult since categorising it would compromise these attributes. In this chapter on sustaining institutional mobile learning innovation, we attempt to embrace all aspects of mobile learning from variation in the type of device and portability to learning formally and informally. We do this by focusing on the people who use the devices rather than on the device(s) and associated technical requirements or contexts.

Sustaining mobile learning: key challenges

In categorising the challenges to the sustainability of mobile learning innovations in educational institutions, we draw on the five components reported in Ng and Nicholas (2013). Four of the components that are based on Cisler's (n.d.) framework for sustaining technology in education are economic, technological, social and political sustainability. We added a fifth component: pedagogical sustainability. In the sections that follow, we describe these components in more detail with a focus on the human interactions and decision-making that are necessary to sustain mobile learning programs in schools or higher education.

Economic sustainability: financial considerations

Economic sustainability refers to the financial capability of an educational institution to ensure the continuity of technology-enabled programs over a long period of time. Many of the reports assert that a major hurdle in the sustainability of many mobile learning projects is overcoming the cost associated with mobile hardware, software, connection, usage charges and technical support. There are also additional costs related to the maintenance of the infrastructure and professional development of staff. Cisler (n.d.) indicated that it is necessary to develop multiple channels of funding, including through partnerships with the community. Community participation connects economic sustainability with the social and political aspects of sustainability, but brings its own challenges.

Technological sustainability: infrastructure, security and devices

In many of the mobile learning reports, *infrastructure* is the most important consideration in adopting these programs in educational institutions. A robust wireless network with high-speed broadband connection and sufficient wireless coverage is essential to handle the high volume of (different) personal devices searching for online access at any one time during the school day or in lectures/tutorial classes in higher education. There need to be sufficient storage, sufficient bandwidth to handle multimedia applications and sufficiently powerful servers to manage multiple devices. In connecting back to economic sustainability, planning needs to consider future network growth and budget for it, as new applications will demand increased bandwidth.

Security and access are issues that have competing priorities, creating different sustainability issues. With the increasing range of personal mobile devices operating on different platforms, there is a need to make use of platform-independent tools. As web-based applications work on most platforms and are able to accommodate common software needs such as multimedia presentations, the institution needs to have a presence in the 'cloud'. This offers advantages in access, but simultaneously challenges in making that access both robust and secure. Cloud-based online storage, sharing and delivery of files and applications through a single sign-in on any device, solves file compatibility issues, where students only require a current version of a browser such as *Safari*, *Chrome*, *Firefox* or *Internet Explorer*, and an Internet connection to access all their files and to navigate the Internet safely. Cloud-based services also enable teachers and students to access their institutional email remotely using their smart mobile device, allow students to access the same course content materials from inside and outside the institution and collaborate with peers in shared online spaces. There are also options for connecting other student data systems such as the ePortal or the virtual learning management environment to the cloud services so that there are fewer security risks as the students' sign-in processes are governed by the same username and password inside and outside the institution. *Google Apps*, *Evernote* and *Microsoft Office 365* are popular cloud services currently used in schools. While there are advantages in access, *security* is the flipside challenge for mobile learning. This includes data protection, protection from viruses on the network and for safety on the Internet. The potential of exposure of sensitive institutional information is increased by permitting students to connect their personal mobile devices to the institutional network while at the same time increasing access.

Decision-making in the setting up of the infrastructure and ensuring security for mobile learning requires the leadership team (head of institution and program coordinators) and IT teams to work together to plan and decide on the needs and goals of the institution and implement what is financially possible. An understanding of how teachers and students will use the mobile devices is also necessary in the planning since they are the major users of the system that is set up. How much and who provides IT support in mobile learning programs need also to be considered carefully. Questions centred on the people carrying out the roles that need to be addressed include: are the infrastructure set-up and security-based tasks going to be outsourced? Is it better to seek consultants' help or to deploy staff within the institution to manage these tasks? Is there sufficient staff expertise and budget to design and manage the setup of an extensive wireless network which allows for capacity to grow and future expansion as the number of mobile devices and apps increase, hence demanding more bandwidth? Should there be separate networks for teachers and students? What are the requirements of users (teachers and students) during the day and outside the classroom?

Social sustainability: community engagement

Social sustainability addresses the involvement of the wider community such as parents, institutional board members, researchers, hardware and software suppliers and/or business partners and the contribution of these stakeholders to the start-up, duration and/or growth of the mobile learning innovation, but it can also involve consideration of the use of mobile devices as part of students' social networks, where the security issues can focus on people rather than data.

Political sustainability: leadership, equity and policy

This refers to the role of leadership and institutional policies required to adopt and maintain mobile learning programs. In bringing new technologies and new programs into the institution, the leaders (heads of schools/faculties and coordinators of mobile learning) need to be able to identify the requirements for successful adoption that is sustainable and manage the change process, for example ensuring teacher readiness and that different learning styles or learner features (e.g. learning difficulties or physical challenges) are considered.

A policy that clearly articulates ownership, responsibility, safeguarding and support for a mobile learning program is necessary but can be challenging to construct. The policy should clearly:

- set out the goals of the program;
- provide a framework for the kinds of learning activities that will be needed;
- specify ways in which teachers will be supported to develop overall pedagogy and specific activities for use in teaching;
- reflect equity associated with ownership and use of mobile devices by including information on the provision of loan, co-contribution or lease-to-buy devices for families that cannot afford to purchase their own devices and/or software;
- specify the device model or alternatively the minimum specification of devices if choice is to be provided to the students and their families;
- include information on accessing the institution's wireless network, who has access to such information under which circumstances and from where;
- state information on the type of technical support available, including provision for charging the mobile devices – do students bringing their own mobile device have to support the device and apps for learning themselves?;

- communicate the guidelines for the use of personal devices within the institution, for example outlining where and when the devices may be used, outlining responsibilities for bringing fully charged devices to classes etc.;
- address digital citizenship such as online etiquette; copyright; guidelines that respect and protect the privacy of others by using only assigned accounts, and viewing and using data to which they are authorised;
- provide information that ensures security for people and online safeguarding, for example in social networking and messaging; refrain from distributing private information about themselves and others
- inform students of their rights and responsibilities, including the use of the devices appropriately, any need to abide by BYOD agreements and consequences of not doing so;
- inform parents and students that the responsibility lies with the students to look after their own devices and that the institution is not liable for loss or damage.

Drawing up an institutional policy for mobile learning will need input and agreement from the whole community – institutional head, coordinator or manager of the mobile learning program, IT staff, teachers, students and in the case of schools, parents.

Pedagogical sustainability: teaching and learning

This refers to teaching/learning practices that support the long-term goals of the mobile learning programs. Pedagogical sustainability defines the roles of teachers and learners in facilitating learning with mobile devices, the preparation and practices that go with it and the nature of peer collegiality required to ensure the best pedagogical practices. It also includes the non-formal and informal learning that mobile devices facilitate and how this connects with the formal aspects of learning with mobile devices.

Teachers are key players in the sustainability of any kind of mobile learning program. They, as well as their students, need to see the value and purpose of using mobile devices and apps in their teaching and learning. By this we mean that teachers need to be able to see better learning outcomes being supported by these devices. For the teacher, it is important that preparing lessons supported by mobile technology and apps does not impose a lot more time or that they can see a longer-term gain in use of time from an initial investment of more time.

There are not many teachers who are able to handle heterogeneous mobile devices with multiple platforms and varied capabilities, although some would argue that the use of browser-based applications and Web 2.0 technologies will solve the problem. However, teachers do not need to know everything about all possible technologies. Rather they need deep knowledge of a relevant selection of tools (Ng, 2015). All smart mobile devices are enabling with respect to consumption activities such as Internet searching, reading text on the screen, watching videos, placing short comments in discussion forums and blogs, and participating in polling exercises, multiple choice quizzes, games, simulations and production activities such as making audio or video recordings, taking photographs or sending messages in various formats. However, basic devices may not have the capabilities to do creative and more complex problem solving tasks. Recognising this means that teachers need to understand the potential and limitations of each of the devices that their students have access to or plan in ways that limitations within the group are minimised and (group) affordances maximised. Table 1.2 is an adaption of Dixon and Tierney's (2012, p. 5) mapping of the computing capability and pedagogical potential of mobile devices. It shows the limited capabilities of a smartphone in comparison to a tablet PC that is also capable of scribing with a stylus. Although smartphones and tablets have their advantages, such as anytime anywhere learning and on-demand communication, there are limitations to the features and usability of these devices. Small screen size (or limited screen resolution), Internet browser speed, battery life, software availability, and support for certain multimedia files are just a few of the considerations. While there are more apps appearing in mobile stores for smartphones, they may not be applicable across the range of mobile devices in a BYOD classroom. The challenge for teachers is twofold: (1) keeping abreast of resources that are pedagogically sound and would impact positively on learning outcomes; and (2) being able to integrate innovatively the resources

Table 1.2 Computing capability taxonomy (adapted from Dixon & Tierney, 2012, p. 5)

Sample capabilities	*Smartphone (android, iOS)*	*Tablet (android, iOS)*	*Laptop (PC, Mac)*	*Tablet PC with pen*
Internet search	X	X	X	X
Voice, video and audio recording	X	X	X	X
Conferencing and collaboration	X	X	X	X
Supports small amounts of typing	X	X	X	X
Screencast video capture	X	X	X	X
Video and audio editing	X	X	X	X
Supports typing of longer assignments	With keyboard	With keyboard	X	X
Multitasks for complex research and knowledge building			X	X
Supports software for CAD, web and graphic design			X	X
Supports programming and handwriting recognition for Maths, Music, Chemistry and Asian characters				X
Note taking with digital pen				X

into the students' learning so that the devices can be used for scaffolding and assessing the learning. An example for the latter would be encouraging students to work collaboratively in groups around the devices that have the capability to access the recommended app. The teacher would still need to articulate the purpose and value of the activity and assess the impact on each individual student's learning. Students collaborating in teams around a device would require a way of storing, retrieving and sharing the data collected on one device, for example when working in teams on a field trip. Cloud storage facilities such as *Dropbox* or *GoogleDrive* are possible means of sharing the data, but teachers will need to be supported to learn how to manage such applications in ways that are accessible to the students but (as appropriate) protect the students from outside observation and do not incur unintended costs.

Teaching with mobile devices is challenging because it requires the teachers to rethink and restructure their educational approach to integrate/adapt the capacities of mobile devices into good pedagogic practices. It requires the teacher to understand how students learn and communicate with mobile technology and to apply appropriate learning theories in the design of lesson plans and assessment. The teacher is required to seek out appropriate mobile apps for teaching specific topics and select appropriate Web 2.0 technologies (i.e. browser-based) to assist students with learning. (S)he needs to have a sufficient level of mobile learning literacy, a subset of digital literacy (Ng, 2013), to make use of mobile devices and tools to teach comfortably without having too heavy a cognitive load that will require him/her to split his/her attention from the teaching, for example solving technical problems associated with the device or software while teaching. In fact, the more the teacher knows about the capabilities of the mobile devices and apps the students are using, the easier it is for the teacher to integrate mobile learning into his/her teaching. It is imperative that the teacher has time to explore mobile devices and new apps, to talk with colleagues about pedagogical applications and/or to observe their classes, to attend professional learning workshops and/or read and participate online in order to keep abreast of changes, new apps and pedagogy. This is not easy and does not happen overnight. Sustaining this kind of activity is a key challenge for educational institutions. Having a mentor for teachers who are early adopters of mobile learning would assist greatly with the teacher's professional development and practice in the classroom. For successful and sustainable mobile learning programs to occur, it is necessary to address both the needs of the teacher and those of their students.

Students are often touted as the main reason for using technology in classrooms. There is general agreement that a mobile learning program such as 1:1 or BYOD where students have their own computing devices enables student-centred pedagogy and motivates student engagement with learning. They are claimed to be able to take account of the multimodal nature of communication (Jewitt, 2008; Nicholas & Starks, 2014).

However, precisely because their affordances are so rich and overlapping it is difficult to identify direct relationships between their use and learning. It is difficult to measure the impact of technology on learning due largely to the difficult nature of collecting research evidence that shows improved performance due solely to the technology and not other environmental factors, such as the teachers' scaffolding techniques or the social interactions that students engage in with the teacher or peers.

Nevertheless, it is possible to gain some insights by attending to the voices of students. In a 2013 study into Australian adolescents' literacy and practices with mobile technology, we asked 12–15 year old (Years 7 to 10) students about their views on learning with mobile technology (Ng & Nicholas, manuscript in preparation). Their responses indicated that in many cases the use of mobile technologies was viewed positively, but not overwhelmingly so. In other words, the preconditions for the use of mobile devices exist, but the students themselves do not see the devices as unambiguously positive educational tools. A total of 1,129 students from seven schools in the Australian states of New South Wales and Victoria participated in the research, comprising 343 year 7 students (12 years old), 319 year 8 students (13 years old), 245 year 9 students (14 years old) and 222 year 10 students (15 years old). The schools represented a spread of single-sex and co-educational schools as well as government and independent schools. About two thirds of the students owned a smartphone and more than 70% of the students were in a laptop or iPad program at school. The Likert-scale statements on learning with mobile technology are shown in Table 1.3. The scale for the statements ranged from 1 to 7 (1=strongly disagree; 7=strongly agree). The internal consistency coefficient of the responses, Cronbach alpha is .885 indicating strong reliability. The results show that the mean values for most of the statements are above 4 (the neutral value) but less than 5 (slightly agree value), indicating that the students were generally positive about learning with mobile technology but not overly enthusiastic. The standard deviations for most of the means are fairly large indicating substantial differences in opinions between the students. The students overall agree strongly about their own abilities to easily learn to work with new technology (mean = 5.88, SD = 1.42) and disagree slightly that learning with mobile devices is frustrating (mean=3.84, SD = 1.88). They agree to some extent that mobile devices, especially mobile phones are distracting and disruptive in the classroom (mean = 4.60, SD = 1.99) and disagree only slightly that mobile devices should not be allowed in the classroom (mean = 3.92, SD = 2.13). These statements indicate that regardless of the policy, the presence of mobile devices alone will be insufficient to sustain a mobile learning program. This means that the burden of the sustaining of such a program falls on the people who are implementing it – the teachers and those who support them.

The Australian study also found that the students were not as attached to their mobile phones as is sometimes suggested. This is evidenced by the

Table 1.3 Mean and standard deviations of responses to statements about learning with mobile technology

Learning with mobile technology	*Mean*	*SD*
I learn new technologies easily	5.88	1.42
Mobile devices, especially mobile phones, are distracting and disruptive in the classroom	4.60	1.99
Mobile devices should not be allowed in the classroom	3.92	2.13
Teachers should teach their students to use their mobile devices sensibly and for learning in the classroom	4.95	1.90
Teachers should discuss and negotiate mobile learning opportunities with their students	4.96	1.77
Mobile devices are good enough to do school work and assignments, e.g. using the word processor or spreadsheet in the mobile device	4.49	2.06
I am interested in learning with mobile devices	4.97	1.84
Mobile devices are good for helping me connect learning that I do at school and learning that I do in other places	4.99	1.73
Mobile devices enable me to be a more self-directed and independent learner	4.91	1.75
Teachers in schools should use more mobile-enabled technology for learning and teaching	4.84	1.86
I am/would be more motivated to learn with mobile technology	4.87	1.86
I (would) listen to podcasts (audio recorded lessons or explanations of concepts) on my mobile device if my teachers produce and make them available to the students	4.72	1.91
I (would) watch videos of lessons or explanations of concepts that my teachers produce for mobile devices	5.00	1.80
I am/would be interested to learn with educational digital games on mobile devices	5.00	1.83
It is important for me to learn more about integrating mobile devices and apps in my learning	4.75	1.75
I would like more opportunities to use my mobile device(s) in my learning	4.99	1.83
I think learning with mobile devices can be frustrating	3.84	1.88

Scale: 1=strongly disagree, 2=disagree, 3=slightly disagree, 4=neutral, 5=slightly agree, 6=agree and 7=strongly agree. (Cronbach alpha=.885)

responses to the statements "I feel lost without my mobile phone" with mean = 4.13, and a standard deviation of 2.10, and "I feel anxious/depressed if nobody sends me an SMS/text message, rings or write messages e.g. on *Facebook* that I can access with my mobile phone each day" (mean 2.74, SD = 1.94). The response to the latter statement indicates fairly strong disagreement with the need to have text or phone messages on a daily basis. This is confirmed in interviews conducted with more than 50 Year 9/10 students in laptop or iPad programs to elicit their views on technology and learning where a comment was "we are not as dependent on it as people think we are". The research found that these 14–15 year old students were more critical in their analysis of how mobile technology is used in educational settings.

They indicated that it was sometimes difficult to see the value and purpose of using technology in classrooms. There was also a general theme that centred on the role of the teacher as being able to "hold their attention", scaffold their learning and explain things well, with one student commenting "I would rather see better teachers". Many students expressed preferences for books and writing as they believed that they learn more when they write by hand (see Mueller & Oppenheimer, 2014 for related findings with lecture note taking).

Similar findings highlighting the influence of factors other than technology were reported by Hallissy et al. (n.d.) on the evaluation of a tablet program implementation in three Irish schools in 2013, and Ipsos MORI (2007) on the JISC project that investigated British school students' use of and attitudes towards new technologies. Like their Australian counterparts, the students had similar views about the educational value of technology and saw the input of the teacher as still very important.

With respect to preferences by tertiary students in the use of technology for their learning, Kennedy et al.'s (2008) report showed that higher education students relied on core technologies for the fairly traditional purposes of communicating and information gathering. There was little use of new technologies such as social networking, file sharing, podcasting and social bookmarking with the students. The Educause Centre for Applied Research's (ECAR) study (Salaway et al.) found that nearly 60% of the students were looking for only moderate amounts of IT in their academic courses. Students also indicated a preference for social networking to be maintained for their private lives (Salaway et al., 2007).

The novelty of digital technology in the classroom is disappearing. While there are expectations by students (and parents) that digital technology is used for learning in the classroom, they also question the value of these technologies in helping them learn better. For example, they are open to books and writing over typing and reading from the screen if those analogue technologies help them learn and remember things better. They do not see technology replacing the teacher. The teachers we interviewed noted similar expectations from their students:

> The thing is there has to be some sort of value or meaning to what they're doing. So if you just put them on there to watch a tutorial they'll just click through it as quickly as they can.
>
> If they're using IT, they have the expectation of being active, rather than passive, they have the expectation of doing something and making something and they tend to have really low tolerance of just sitting there and watching something or reading something and doing something that's really passive and absorbing, as opposed to doing something or making something or modifying something, something interactive.
>
> I found that some students were: "I'm sick of my iPad", "I don't want to do it anymore", "it's difficult". They were trying to do things

> like graphing experimental results on the iPad, and they'd actually prefer to draw it on paper.

Sustaining a practice will be much less difficult if the people involved see it as worthwhile. Hence for pedagogical sustainability of learning with mobile technology programs, the voice of the students is important and teachers should elicit their views on a regular basis, obtain feedback and allow for opportunities to negotiate how and what they could learn with mobile devices and apps. Tasks conducted on mobile devices need to be meaningful and relevant. In other words, it is the people who create the learning activities and the resources that support them who are the crucial players in sustaining innovation in learning with mobile devices.

In the next section we will discuss the challenges that need to be considered in implementing mobile learning programs, in particular BYOD. We will first discuss the various models of BYOD that could be implemented in educational institutions and the implications associated with each of the models, including the challenges. As will be seen in the next section, BYOD models are device-centred, which is inadequate for our people-centred framework. Following the description of the various models for BYOD, we will describe in more detail the key challenges that BYOD learning faces in educational institutions, with a focus on the people whose roles are underpinned by purpose and values necessary to bring about the transformation for sustainability.

Bring-your-own-device (BYOD) and its challenges

Bring your own device (BYOD) refers to a model of technology-mediated learning where students bring a personally-owned smart mobile device from home to school for the purpose of learning. The device is usually a laptop, netbook, smartphone or tablet computer, containing apps for learning and is able to connect to the Internet to access information anywhere in the educational institution. The capabilities of mobile devices outlined in Table 1.1 enable students to: use them as ebook readers that could display textbooks containing visual content and interactive elements; create audio or video resources as part of or evidence of their own learning; access information and video content via the web browsers and video players; communicate socially on social networks via blogs or wikis or in discussion groups or in real-time through video conferencing tools and explore and exploit the range of apps available for these devices. As Johnson et al. stated in the 2013 Horizon Report:

> Learning institutions all over the world are exploring ways to make their websites, educational materials, resources and opportunities all available online and optimised for mobile devices. The most compelling facet of mobile learning right now is mobile apps. Smartphones

> and tablets have redefined what we mean by mobile computing, and in the past four to five years, apps have become a hotbed of development, resulting in a plethora of learning and productivity apps. These tools, ranging from annotation and mind-mapping apps to apps that allow users to explore outer space or get an in-depth look at complex chemicals, enable users to learn and experience new concepts wherever they are, often across multiple devices… Tablets, smartphones and apps have become too capable, too ubiquitous and too useful to ignore.
>
> (Johnson et al., 2013, pp. 16–17)

At the institutional level, a reason to adopt BYOD is the view that budgetary pressures on one-to-one learning could be eased by capitalising on the devices that students already own and are familiar with (Johnson et al., 2013; Lee, 2013). For example, in Australia, the government-funded *Digital Education Revolution* initiative (Australian Information and Communications Technology in Education Committee [AICTEC], 2009) to ensure that all year 9–12 students had access to a laptop is ending. Many schools are now exploring the potential of BYOD schemes (e.g. St Joseph's Nudgee College in Brisbane, http://www.nudgee.com/byod) where in some schools, it is recommended that a generic tablet (e.g. iPad or Lenovo tablet) can be bought for students as a personal device and brought to school to support their learning. Others outline the technical requirements and allow parents to choose a device that matches them (e.g. Sydney's Chatswood High School, http://www.chatswoodhighschool.com/content/byod-information).

BYOD is one form of learning with mobile devices, perhaps the form that brings its characteristics and affordances most sharply into focus. Education institutions facing budgetary pressures embrace BYOD by shifting the cost of the digital devices on to the students and their families. This shift in costs can make the innovation much more economically sustainable for the school, but such an approach can create inequity within the classroom and between schools. The innovation may not be pedagogically sustainable if some students are able to use the most up-to-date devices while others will have older devices with less functional technology. It is crucial that teachers are supported to plan and teach in ways that students unable to afford a smart mobile device or one that is insufficiently sophisticated to carry out the tasks set by the teacher have access to a means of doing so. One strategy could be a loan from the institution, which has to be budgeted for, but a longer-term strategy will involve designing learning activities where students work together and learn to share their devices.

Models of BYOD

In his literature review on BYOD, Stavert (2013) described the different models for BYOD proposed by Sweeney (2012), Dixon and Tierney (2012) and the Alberta Guide (Alberta Education, 2012). The models all share key features such school-defined standard devices to students' personal smart devices (with different platforms and different functionalities but with access to the Internet). In this section, we summarise these models by adapting Dixon and Tierney's (2012) five-models of BYOD, integrating other researchers' models and highlighting the challenges associated with each model.

1 *School-defined single platform laptop or tablet* where the school defines the required minimum specifications for student laptop or tablet. The standard model ensures that all students have access to the same capabilities and makes the planning of learning activities by the teacher easier. A standard device model for a school enables volume-purchasing power, lowering the cost of the device for parents. It also simplifies IT management of the devices.
2 *School-defined single platform laptop, plus another device* where students are allowed to bring their own personal device such as the smartphone, on top of having a common standard device in the class, as in 1 above. This allows students to legitimately bring their personal device and use it as a supplement. The maintenance of the additional device is the responsibility of the student. Teachers plan lessons around the school-defined standard device.
3 *School-defined multi-platform laptops* where as long as the student's devices adhere to minimum specifications, different platforms and brands are acceptable. While this provides choices for parents and students who may have preference for a type of platform/manufacturer, it creates more work for IT staff to manage a variety of laptops since technical support staff are required to be familiar with several platforms. The purchasing power and bulk discounts for software and hardware are reduced and both lesson planning and teaching need more consideration as some programs may not be available across platforms or may operate differently.
4 *Student-choice of laptop or tablet* where the students bring a device of choice, including netbook or a computer tablet, purchased fully by the parents. Similar to 3 above, the model provides choice for parents and students with preference for a type of platform or brand. The challenges are more pronounced in this model where the different capabilities of the devices may limit either consumption and/or production, particularly in creative activities. The purchasing power, bulk discount purchasing and licensing options are reduced. Technical support staff need to be familiar with a variety of platforms across the different devices, for example Windows, Android, Apple and iOS platforms. Teachers need to be familiar with the variety of platforms and capabilities of the different devices, including having some knowledge of apps that are available across all the platforms and how they will function in the different devices. Pedagogically, variation in the availability,

operation or cost of applications could limit what teachers can do with the varied devices, apart from Internet search, reading text and watching videos.

5 *Bring anything providing that it connects to the Internet* where there is no specification for the device (such as screen size, storage or ports) and students bring can any device from home as long as there is Internet connectivity capability. These devices could be a smartphone, eReader, netbook, notebook or tablet. The challenges in this model are similar to those in 4 above but more pronounced due to the inclusion of a larger variety of devices with different functionalities, adding further to the complexity of IT management and teacher's lesson planning. Contradictory to this perception is the view that implementing this model of BYOD is easier on the IT department because students manage their own devices (Williams, 2012). Other researchers have discussed the emergence of peer-to-peer support that shifts the technical troubleshooting and support to students (Brown & Diaz, 2010; Inman, 2012; Sweeney, 2012) where they help one another manage their devices.

Model 5 is the model that people usually mean when referring to BYOD in broad terms. In higher education, this will almost certainly be the case, although some faculties have gone with model 1, for example, the Science Faculty at the University of South Australia provides each of their undergraduate students with an iPad in their first year of enrolment (see http://www.adelaide.edu.au/news/news41241.html). Hence, our reference to BYOD will be based on this model where hybrid mobile devices with Internet access are used. In order to design a conceptual framework for sustaining mobile learning in educational institutions, it is necessary to first discuss the key challenges of implementing mobile learning.

Studies have identified many challenges associated with implementing and sustaining BYOD innovation and more generally, mobile learning in schools (Butterman, 2012; Dixon & Tierney, 2012; Hallissy et al., n.d.; Raths, 2012; Stavert, 2013) and in higher education (Ackerman & Krupp, 2012; Brown & Diaz, 2010; Santos, 2013). BYOD involves the use of mobile technology for learning but has the additional characteristic of variation in both hardware and software. As a result it is more difficult to form a clear view of either a policy framework or appropriate practices.

While peer-reviewed journal articles on the implementation and the degree of success of mobile learning or BYOD at the institutional level are still scarce in the literature, there are a number of reports that describe the implementation process (or aspects of the process) and the associated challenges. A few examples of school-based reports are the Oak Hills Local School District in Ohio report (Vander Veen & Cole, 2012); the 2013 research study on the use of tablet devices in Irish schools (Hallissy et al., n.d.); a report on the technical aspect of BYOD in four US-based schools (Raths, 2012) and a report on the adoption of mobile handheld technologies in ten Australian schools (Pegrum, Oakley, & Faulkner, 2013). In higher education, the challenges of BYOD reported in several studies include: Ackerman and Krupp's (2012) identification of the five challenges of BYOD as security, stakeholders, policies, professional development, and financial planning; Santos' (2013) report on challenges as network infrastructure, network security, IT support, equity issues and BYOD disruptions, and Nykvist's (2012) report on access to common software, data storage, retrieval and presentation, network infrastructure and multiple platform familiarity. These and other reports form the basis for the challenges that we will describe in the next section on the implementation and sustaining of mobile learning in education. The other reports include Butterman's (2012) article on planning for BYOD, Williams' (2012) article on managing BYOD effectively, diFilipo's (2013) consideration of BOYD policy for higher education, Brown and Diaz's (2010) *Educause Learning Initiative* white paper report on mobile learning and Sweeney's (2012) report for Australia and New Zealand on conversations with school and higher education stakeholders about successful BOYD decision making.

Tensions in relation to sustainability arising from decisions to adopt BYOD reflect the rise in the number of students bringing their own personal devices into the classroom for purposes other than mandated learning, with these devices often having greater capabilities than the ones provided by the schools (Sweeney, 2012). As mobile devices are an inseparable part of students' lives, by allowing students to access the same devices at school the potential is created for learning to be extended to times and places outside the classroom. This approach creates opportunities for continuity in learning. However, the fact that students' own devices have greater capacities than those that the teachers and institutions may be ready for, can contribute to lack of student interest if advantage is not taken of these greater capacities. Nevertheless, policies that once were intended to ban mobile devices are being revised to BYOD policies by Departments of Education and educational institutions. As Alberta Education (2012, p. 4) stated:

> The reality is that web-based tools and resources have changed the landscape of learning. Students now have at their fingertips unlimited access to digital content, resources, experts, databases and communities of interest. By effectively leveraging such resources, school authorities not only have the opportunity to deepen student learning, but they can also develop digital literacy, fluency and citizenship in students that will prepare them for the high tech world in which they will live, learn and work.

BYOD approaches have been designed to improve access to the affordances of mobile learning. As indicated earlier, the tensions between access and security can mean that the access imperative of BYOD creates major issues for technological and social sustainability. A key issue here is the security of the wireless network and the data that is accessed and distributed on it.

Such a security risk could result from the loss of a mobile device that is unprotected. Such a loss could mean exposure of personal or institutional data that the student might have stored in that device. A step towards protecting the wireless infrastructure for a BYOD program is to provide a student network that is separate from the one used by teachers and administrators, but this increases the challenges for the technical design of the institutional infrastructure. This creates distinct domains for content, but does not necessarily enable discrimination between different kinds of content. The design would need to include authentication measures that enable the monitoring of Internet usage while ensuring that only legitimate users are allowed to access the network.

While a web filter can provide Internet access controls, it may block potential useful content as well as reducing the speed in accessing the online materials. It is also possible that mobile devices can communicate with various networks concurrently, regardless of firewalls, and if a mobile device is connected to an institution's network as well as a public network, an unprotected path to the institution's central information system could be potentially opened, creating a security issue (Santos, 2013). The diversity of mobile devices is such that Markeelj and Bernik (2012) asserted that there is no available system that facilitates the monitoring of an institution's information technology (IT) system with respect to the accessing and transferring of data via mobile devices. One way to address some aspects of this issue is to ensure that all students register their mobile devices with the IT department so that their usage could be monitored, but again this provides sustainability issues for IT staff, particularly as students change their devices. The above strategy can be supported by being very explicit in the institution's digital technology or BYOD policy about security risks and educate students (and staff) about responsibilities as users in security practices and network policies, but this requires additional planning, co-ordination and oversight to keep the information current and may have implications for staffing levels or processes and cost.

In school contexts involving BYOD, it is crucial to communicate clearly the plans for purchase of mobile device by parents themselves and the purposes, values, costs and goals in using these devices in the classroom. As with any new innovations, there will be supporters and non-supporters among the parents and teachers. The final decision on the direction of the innovation and the choice of mobile device(s) for the school as well as processes for review need to be clearly outlined and communicated, including policies on use and support and how issues of damage, distractions and thefts will be dealt with. Other interactions with the wider community include negotiations with companies that supply the hardware and software, getting the endorsement and support of district or state Departments of Education, working with academics and professional organisations to provide professional development for teachers, working with researchers from universities to undertake research and evaluation, talking with teachers to elicit professional development needs related to mobile learning programs and finding out where the students are with respect to the use of mobile devices for learning. It is important to elicit the attitudes of the users (teachers and students) and the sole- or co-contributors to the financial sustainability of the program, the parents. A line of communication between these stakeholders needs to be permanently open and relevant staff responsive so that issues are heard and needs addressed in order to sustain the mobile learning program. While there are many sustainability issues, a key feature of all of them is that people are involved. We turn to that issue in outlining our model of sustainability for learning with mobile devices.

The person-centred sustainable model for mobile learning

The five components of sustainability discussed above shape our theoretical approach to sustainable mobile learning. The approach acknowledges the complex relationships between the technical aspects and people as well as the interpersonal relationships between management (institutional heads and program managers), teachers, students, technicians and the wider community (parents, suppliers, policy-makers, software developers and researchers). Sustaining a new technology-based initiative is a complex process. Al Neimat (2005), citing Tilmann and Weinberger (2004), stated that the success of IT projects in business is dependent on the process of project management and its alignment with the culture of the organisation. Smith (2008), in agreement with Al Neimat's argument, stated that IT projects fail not because of the technology but because of the people. He further indicated that the success or failure of a project depends on the team dynamics, the skills and effectiveness of the people involved, their openness to change and their ability to focus so that there needs to be a focus on people first. Hence, considering the human factors upfront and how the stakeholders interact with each other and with the technology is important in theorising sustainable mobile learning programs. For this reason, we propose a person-centred model for sustainable mobile learning as shown in Figure 1.1, which is adapted from Ng and Nicholas (2013, p. 699). The non-hierarchical, person-centred model shows the relationships between players in a mobile learning program and the levels of intersection (interactions) between them and with the technology. These interactions require communication (negotiation, consultation, feedback), support and trust between stakeholders.

The person-centred model, shown in Figure 1.1, indicates that teachers are central to the success and sustainability of mobile learning, as shown, by the largest number of direct interactions with other people involved – leadership and management team, parents, students and technician. It is essential that leaders consult with and provide (technical, professional and emotional) support to teachers. Leaders need to work with the teachers to create a trusting learning environment with the students and incorporate the informal learning with mobile devices that the students would bring

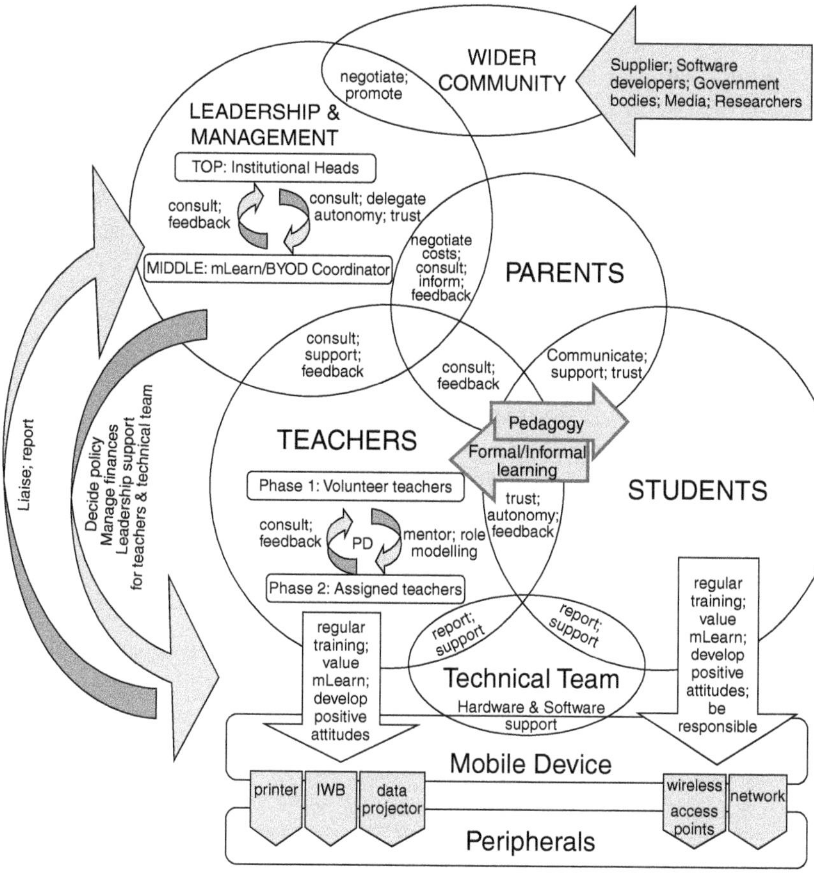

Figure 1.1 Person-centred sustainable model for mobile learning (adapted from Ng & Nicholas, 2013, p. 699).

into the classroom. At a technical level, teachers are required to develop a good relationship with the mobile devices so that they are comfortable with and develop positive attitudes toward mobile devices and their affordances, valuing mobile learning. This requires experience, room to play and experiment and the development of practices that support classroom experimentation. Similarly, the students need to learn to value learning with mobile devices and take responsibility for the care of the devices and learn how to use the devices to participate genuinely in formal learning. Indeed, this is a prime opportunity for learners to see themselves as positively contributing to the shaping of future teaching.

Technology provides the means to host easily searchable information, it provides the platform for apps to be functional and for creative and meaningful activities to take place. But it is teachers who teach with them and students who learn with them and hence, together, they are crucial to sustaining its existence. To sustain innovative mobile learning programs, stakeholders need to carry out their roles responsibly and in a timely fashion, for example the IT person in ensuring that the infrastructure is sufficiently robust to handle the heavy data traffic when several classes use mobile technology at the same time. Focusing on the people involved means providing the necessary support for them to carry out their roles with minimal obstacles and frustrations. It is in the successful interactions between the key players that the economic, political, social, technical and pedagogical elements of sustainability presented in the preceding section will be satisfied.

References

Ackerman, A.S., & Krupp, M.L. (2012). Five components to consider for BYOT/BYOD. *Proceedings of the IADIS International Conference on Cognition and Exploratory Learning in Digital Age* (CELDA 2012), pp. 35–41. Retrieved May 8, 2014 from http://files.eric.ed.gov/fulltext/ED542652.pdf

AICTEC (2009). *Digital education revolution implementation roadmap*. Retrieved March 24, 2014 from http://ehlt.flinders.edu.au/education/eduwiki/lib/exe/fetch.php?media=students10:research_docs:der_implementation_roadmap.pdf

Alberta Education. (2012). *Bring your own device: A guide for schools*. Edmonton: Alberta Education.

Al Neimat, T. (2005). *Why IT projects fail*. The Project Perfect White Paper Collection. Retrieved December 8, 2011 from http://www.projectperfect.com.au/downloads/Info/info_it_projects_fail.pdf

BBC (2008). *Mobile internet usage on the rise*. Retrieved April 8, 2014 from http://news.bbc.co.uk/2/hi/technology/7748372.stm

Brown, M., & Diaz, V. (2010). Mobile learning: Context and prospects. *A report on the ELI Focus Session: Educause Learning Initiative*. Retrieved May 7, 2014 from http://net.educause.edu/ir/library/pdf/ELI3022.pdf

Butterman, E. (2012). Planning for BYOD. *Scholastic Administrator*, *12*(1), 55–57.

Cerwall, P. (2012). *Report shows rapid smartphone uptake and doubling of mobile data traffic*. Retrieved April 3, 2014 from http://www.ericsson.com/news/121121-report-shows-rapid-smartphone-uptake-and-doubling-of-mobile-data-traffic_244159017_c

CISCO (2012). *Cisco visual networking index: Global mobile data traffic forecast update*, 2011–2016. Retrieved May 5, 2014 from http://www.cisco.com/en/US/solutions/collateral/ns341/ns525/ns537/ns705/ns827/white_paper_c11-520862.html

Cisler, S. (n.d.). *Planning for sustainability: How to keep your ICT project running* [schools online]. Retrieved August 3, 2011 from http://www2.ctcnet.org/ctc/Cisler/sustain.doc

Clough, G., Jones, A., McAndrew, P., & Scanlon, E. (2009). Informal learning evidence in online communities of mobile device enthusiasts. In M. Ally (Ed.), *Mobile learning: Transforming the delivery of education and training. Issues in distance education* (pp. 99–112). Canada: Athabasca University Press.

diFilipo, S. (2013). *The policy of BYOD: Considerations for higher education*. Retrieved December 4, 2014, from http://www.educause.edu/ero/article/policy-byod-considerations-higher-education

Dixon, B., & Tierney, S. (2012, July 1). *Bring your own device to school.* Retrieved January 28, 2014, from http://www.microsoft.com/education

Georgiev, T., Georgieva, E., & Smrikarov, A. (2004). M-learning: a new stage of e-learning in *Proceedings of International Conference on Computer Systems and Technologies*, Rousse, Bulgaria. doi:10.1145/1050330.1050437

Goodchild, T., & Chenery-Morris, S. (2011). Educational podcasts at University Campus in W. Ng (Ed). *Mobile technologies and handheld devices for ubiquitous learning: Research and pedagogy* (pp. 187–208). Hershey, PA: IGI Global Publishing.

Griffith, C. (2013a). Desktop sales nosediving as tablets take off: Gartner. *The Australian.* Retrieved May 6, 2014 from http://www.theaustralian.com.au/technology/desktop-sales-nosediving-as-tablets-take-off-gartner/story-e6frganx-1226641477479

Griffith, C. (2013b). Huawei's brave phablet, the Ascend Mate. *The Australian.* Retrieved May 19, 2014 from http://www.theaustralian.com.au/technology/personal-tech/huaweis-brave-phablet-theascend-mate/story-e6frgazf-1226641441865

Hallissy, M., Gallagher, A., Ryan, S., & Hurley, J. (n.d.) *The use of tablet devices in ACCS schools.* Retrieved May 13, 2013 from http://www.athycollege.ie/uploads/1/0/2/6/10262974/the_use_of_tablet_devices_in__accs_schools.pdf

Hwang, G.-J., & Tsai, C.-C. (2011). Research trends in mobile and ubiquitous learning: A review of publications in selected journals from 2001–2010. *British Journal of Educational Technology*, *42*(4), E65–E70.

Inman, A. (2012). *How to start a BYOD program.* Retrieved May 2, 2014, from http://www.edtechmagazine.com/k12/article/2012/04/how-start-byod-program

Ipsos MORI. (2007). *Student expectations study: Key findings from online research and discussion evenings held in June 2007 for the Joint Information Systems Committee: JISC.* Retrieved May 19, 2014 from http://www.jisc.ac.uk/media/documents/publications/studentexpectations.pdf

ITU [International Telecommunications Union] (2013). *The world in 2013: ICT facts and figures.* Retrieved May 6, 2014 from http://www.itu.int/en/ITU-D/Statistics/Documents/facts/ICTFactsFigures2013-e.pdf

ITU [International Telecommunications Union] (2014). *The world in 2014: ICT facts and figures.* Retrieved May 6, 2014 from http://www.itu.int/en/ITU-D/Statistics/Documents/facts/ICTFactsFigures2014-e.pdf

Jewitt, C. (2008) Multimodality and literacy in school classrooms. *Review of Research in Education*, *32*(1), 241–267.

Johnson, L., Adams Becker, S., Cummins, M., Estrada, V., Freeman, A., & Ludgate, H. (2013). *NMC Horizon Report: 2013 K-12 edition.* Austin, TX: The New Media Consortium.

Kelly, T., & Minges, M. (2012). World Bank report 2012: Executive summary. *Information and Communications for Development 2012: Maximizing Mobile*, pp. 3–9. Washington, DC: World Bank. Retrieved May 5, 2014 from http://www.worldbank.org/ict/IC4D2012.

Kennedy, G., Judd, T. S., Churchward, A., Gray, K., & Krause, K. (2008). First year students' experiences with technology: Are they really digital natives? 'Questioning the net generation: A collaborative project in Australian higher education', *Australasian Journal of Educational Technology*, *24*(1), 108–122

Kukulska-Hulme, A., Sharples, M., Milrad, M., Arnedillo-Sánchez, I., & Vavoula, G. (2009). Innovation in mobile learning: A European perspective. *International Journal of Mobile and Blended Learning*, *1*(1), 13–35.

Lee, M. (2012). BYOT. *Australian Educational Leader*, *34*, 45–46.

Lomas, N. (2013). *IDC: Tablet sales grew 78.4% YoY in 2012 – expected to pass desktop sales in 2013, portable PCs in 2014*. Retrieved March 8, 2014 from http://techcrunch.com/2013/03/27/idc-tablet-growth-2012-2017/

Looi, C. K., Seow, P., Zhang, B., So, H. J., Chen, W., & Wong, L. H. (2010). Leveraging mobile technology for sustainable seamless learning: a research agenda. *British Journal of Educational Technology*, *41*(2), 154–169.

Markeelj, B., & Bernik, I. (2012). Mobile devices and corporate data security. *International Journal of Education and Information Technologies*, *1*(6), 97–104.

Minges, M. (2012). World Bank report 2012: Overview. *Information and communications for development 2012: Maximizing mobile*, pp. 11–30. Washington, DC: World Bank. Retrieved May 5, 2014 from http://www.worldbank.org/ict/IC4D2012

Mueller, P., & Oppenheimer, D. (2014). The pen is mightier than the keyboard: Advantages of longhand over laptop note taking. *Psychological Science*, *25*(6), 1159–1168.

Mukherjee, A. (2014). Google developing tablet with advanced vision capabilities: report. *The Australian*. Retrieved May 30, 2014 from http://www.theage.com.au/digital-life/tablets/google-developing-tablet-with-advanced-vision-capabilities-report-20140524-zrn0k.html

New South Wales Department of Education and Communities (n.d.). *Student bring your own device policy (BYD)*. Retrieved April 25, 2014 from https://www.det.nsw.edu.au/policies/technology/computers/mobiledevice/PD20130458.shtml

Ng, W. (2013). Conceptualising mlearning literacy. *International Journal of Mobile and Blended Learning*, *5*(1), 1–20.

Ng, W. (2015). *New digital technology in education: Conceptualising professional learning for educators*. New York: Springer.

Ng, W., & Nicholas, H. (2013). A framework for sustainable mobile learning in schools. *British Journal of Educational Technology*, *44*(5), 695–715.

Nicholas, H., & Starks, D. (2014) *Language education and applied linguistics: Bridging the two fields*. London & New York: Routledge.

Nielsen (2010). *Mobile youth around the world*. Retrieved May 5, 2014 from http://www.nielsen.com/us/en/reports/2010/mobile-youth-around-the-world.html

Nykvist, Shaun S. (2012). The trials and tribulations of a BYOD science classroom. In Yu, Shengquan (Ed.) *Proceedings of the 2nd International STEM in Education Conference*, Beijing Normal University, Beijing, China, pp. 331–334.

Ofcom (2013). *Children and parents: Media use and attitudes report*. Retrieved March 8, 2014 from http://stakeholders.ofcom.org.uk/binaries/research/media-literacy/october-2013/research07Oct2013.pdf

Pachler, N. (2009). Research methods in mobile and informal learning: some issues. In G. Vavoula, N. Pachler, & A. Kukulska-Hulme (Eds.), *Researching mobile learning: frameworks, tools and research designs* (pp. 1–15). Oxford: Peter Lang Publishing.

Parsons, D., & Ryu, H. (2006). *A framework for assessing the quality of mobile learning*. Retrieved December 10, 2010 from: http://citeseerx.ist.psu.edu/viewdoc/download?doi=10.1.1.108.2612&rep=rep1&type=pdf

Pegrum, M., Oakley, G., & Faulkner, R. (2013). Schools going mobile: A study of the adoption of mobile handheld technologies in Western Australian independent schools. *Australasian Journal of Educational Technology*, *29*(1), 66–81.

Petrova, K., & Li, C. (2009). Focus and setting in mobile learning research: A review of the literature. *Communications of the IBIMA*, *10*, 219–226.

Pew Internet Research (2012). *Digital differences*. Retrieved March 8, 2014 from http://www.pewinternet.org/files/old-media//Files/Reports/2012/PIP_Digital_differences_041312.pdf

Raths, D. (2012). Are you ready for BYOD? *T.H.E. Journal*, *39*(4), 28–32.

Rideout, V. (2013). *Zero to eight: Children's media use in America 2013*. Retrieved March 10, 2014 from http://www.commonsensemedia.org/research/zero-to-eight-childrens-media-use-in-america-2013

Salaway, G., Caruso, J. B., & Nelson, M. R. (2007). *The ECAR study of undergraduate students and information technology, 2007*. Boulder, CO: Educause Center for Applied Research. Retrieved May 20, 2014 from http://www.educause.edu/ecar

Santos, I. M. (2013). Key challenges associated with bringing personal mobile devices to the classroom. *QScience Proceedings*, (12th World Conference on Mobile and Contextual Learning [mLearn] 2013). Retrieved May 8, 2014 from http://www.qscience.com/doi/pdfplus/10.5339/qproc.2013.mlearn.16

Sharples, M., Taylor, J., & Vavoula, G. (2007). A theory of learning for the mobile age. In R. Andrews & C. Haythornthwaite (Eds.), *The Sage handbook of e-learning research* (pp. 221–247). London, UK: Sage.

Smith, D. (2008). Why do most IT projects fail? It's not because of the technology. *Portland Business Journal*. Retrieved November 1, 2011 from http://www.bizjournals.com/portland/stories/2008/10/20/smallb4.html?page=all

Stavert, B. (2013). *BYOD in Schools Literature Review 2013*. Sydney, Australia: DEC.

Sweeney, J. (2012). *Nine conversations for successful BYOD decision making – A report for Australia and New Zealand*. Retrieved May 8, 2014 from https://marketing.dell.com/Global/FileLib/BYOD_REPORT/Nine_Conversations_On_BYOD_-_IBRS_report_-_Dell.pdf

Tilmann, G., & Weinberger, J. (2004). Technology never fails, but projects can. *Baseline*, *1*, 26, 28.

Traxler, J. (2007). Defining, discussing, and evaluating mobile learning: the moving finger writes and having written. *International Review of Research in Open and Distance Learning*, *8*(2), 1–12.

Vander Veen, Z., & Cole, T. (2012). *Bring your own device*. Ohio: Oak Hills Local School District. Retrieved May 13, 2014 from http://ohlsd.org/portfolio/wp-content/uploads/2012/04/bring_your_own_device.pdf

Whitney, L. (2009). *AdMob: iPhones, Android phones on the rise*. Retrieved April 7, 2014 from http://news.cnet.com/8301-1035_3-10366193-94.htm

Williams, C. (2012). Managing BYOD Effectively. *District Administrator*, *48*(9), 84–85.

World Bank (2012). *Information and communications for development 2012: Maximizing mobile*. Washington, DC: World Bank. Retrieved May 8, 2014 from: http://www.worldbank.org/ict/IC4D2012

Wu, W. H., Jim Wu, Y. C., Chen, C. Y., Kao, H. Y., Lin, C. H., & Huang, S. H. (2012). Review of trends from mobile learning studies: A meta-analysis. *Computers & Education*, *59*(2), 817–827.

2 Sustaining the unsustainable

Reframing global learning for the twenty-first century

Tim Best, Teresa Franklin and Scott Walthour

Introduction

The concept of the university as an ivory tower resonates with many people. Economist Dan Klein shows that the top 35 economics departments worldwide pull 76% of their faculty from their own graduates. He argues that the academic culture is pyramidal, not polycentric, and resembles a closed and genteel social circle. The result is a self-organising and self-validating circle (Klein, 2005).

This self-imposed isolation has caused higher education to be questioned as to how well it delivers on the promise of meaningful learning experiences that engage learners and meet the needs of society. Scholars and policy makers have observed that higher education is not actively participating in the interconnected global nature of the world and are calling for a change. Swail (2002) suggests to his readers "that the rules are changing, and there is an increased pressure on institutions of higher education to evolve, adapt, or desist" (p. 16). How students learn is dramatically changing with the proliferation of Web 2.0 technologies and applications. This will continue to change with the connected features of Web 3.0 technologies (Bransford, Brown, & Cocking, 2000; Spivack, 2014). Web 3.0 will bring a more connected web in which a semantic Web provides distributed databases, natural language processing, machine learning and reasoning, and disruptive environments in which research is lacking and the environment is not fully understood (Matusky, 2014; Spivack, 2014). The need for faculty to merge pedagogy and andragogy to form a more sustainable model of technology use in open and connected learning environments will be critical. Can higher education sustain its ivory tower reputation in a world that evolves with each new technological change?

Acknowledging the opportunities and challenges mobile learning presents to teaching and learning in higher education, the century-old lesson learned combined with a relatively modern model for examining technology may hold important keys for realising its potential and ensuring its sustainability. The lesson emerges by comparing the meagre productivity gains seen when electric and internal combustion engines overtook the steam engine around the turn of the twentieth century with the development and initial use of information technology at the turn of this century. In both cases, business productivity gains were not realised for decades because those managing industry only used advanced technology adaptively, making marginal productivity gains at best (Brynjolfsson & McAfee, 2014). Transformational use of technology that makes exponential growth in learning possible may be a product of new tools *combined* with methods that make use of the affordances of new technologies.

A corollary to this phenomenon is the increasing gap between technical innovation and our ability to use innovations productively and well – as posited in Clayton Christensen's disruptive innovation theory (Christensen, Johnson, & Horn, 2010). It appears the engineers' ability to create new technologies often outstrips our ability to use them well. Given public schools' penchant for attempting to stay current with technology as a way of competing for attention and students in an increasingly competitive environment (Chubb, 2014), what solutions are there for breaking the rigid meta-stable systems in higher education? Typically an outlier, the enlightened tech advocate who often with heroic self-sacrifice creates wonderful one-off learning environments, accomplishes this solution. Often such individuals pour countless hours into improving their craft with technology, doing so for the gains he/she believes are possible if new tools are used creatively as open systems that transform learning. Unfortunately, such a model is not sustainable. The tech heroes burn out because of the tremendous effort required to sustain such creativity. They are promoted out of their present circumstance, or they eventually lower their expectations in the face of unrealistic sustainability challenges. What is needed is a strategy for large-scale change.

Change is difficult for a closed system such as higher education because by its very nature the mission is to maintain the status quo. Moving a rigid meta-stable education system is almost impossible. New ideas or technologies do not have a transformative impact on it. Rather than transforming university culture, the education system absorbs and adapts these elements to meet existing needs and goals. As Franklin notes,

> [e]ducators in general are somewhat resistant to the idea of an open system or world in which a learner can *reach out and touch him/her* which is the direction that mobile learning and cell delivery of content takes the learner in a world in which 24/7 access makes locating experts, in this case faculty, for questions, discussion, and collaboration an easy task. Through the use of their mobile device, our learners can quickly *fact check* a class lecture, *find people* who may know more about the topic than the faculty, *find people* who have completed the assignment last year and ask them for help — or *copy* intellectual property directly and *Facebook* or *tweet* how bad/boring your class is — all while attending your lecture.
>
> (Franklin, 2010, p. 263)

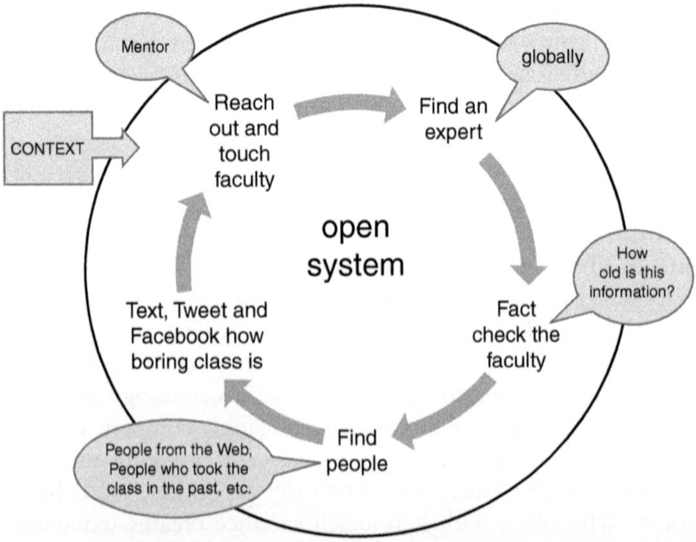

Figure 2.1 The new open system of the classroom (Idea for the graphic from Oblinger, D. (2011, March/April). Complexity, communication and control: Perspectives on mobile. *Educause Review*, *46*(2), 1. The text of this article is licensed under the Creative Commons Attribution-NonCommercial-NoDerivs 3.0 License http://creativecommons.org/licenses/by-nc-nd/3.0/).

In *The Second Machine Age,* Brynjolfsson and McAfee (2014) posit a school of thought that holds that the true work of innovation is not coming up with something big and new, but instead recombining things that already exist. The current global digital culture is a culture of remix and re-mixability, where user-generated content exists both within and outside higher education contexts, and supports as well as subverts faculty control (Jenkins & Deuze, 2008). As more and more learners create user-generated content, there are growing teams of audience activists leveraging the participatory culture of the web. Harnessing collective intelligence, learners are finding ways to move beyond the faculty as 'sage on the stage' to an open system for learning that depends upon context rather than classroom. Questions are answered on a need-to-know basis with learners relying on web-based resources and experts that they contact online. Learners can now expand the participants to include participants within the surrounding culture. The expansion of this media based culture has led to what Yochai Benkler has described as a "'hybrid media ecology' within which commercial, amateur, governmental, nonprofit, educational, activist and other players interact with each other in ever more complex ways. Each of these groups has the power to produce and distribute content and each of these groups is being transformed" (as cited in Jenkins & Deuze, 2008, p. 1).

Given the nature of participatory engagement, each of Chickering and Gamson's (1987) "seven principles for good practices in undergraduate education" often presented to faculty in higher education as a means of improving teaching and learning may actually be well attended to online. Some examples of the seven principles in a connected, interactive participatory culture online might go as follows:

1 *Encouraging frequent student-faculty contact* (Hacker, 2000) – in a participatory culture the learner sends out a question and receives many responses with follow-up often on the decision concerning the question.
2 *Providing timely feedback for students* (Hacker, 2000) – the speed at which a learner with a question can make connections to others on the web is quick and direct. The web creates an environment in which all of the participants can voice agreement and dissent, thus providing the learner with multiple levels of feedback from which to finalise a decision or answer.
3 *Promoting active learning and cooperation among students* (Hacker, 2000) – the interaction of many 'experts' with whom the learner may identify and connect creates an environment in which the learner is actively involved in multiple opportunities to engage with others having experiences with the content being discussed or the ability to connect the learner to others with knowledge. This give and take of building, distributing, and sharing content promotes an environment in which cooperation among the learners guides the direction of the question to a solution or multiple solutions.

These three examples from the *Seven Principles* provide a glimpse into not only the good practices of education, but they also offer an avenue for higher education to succeed in a mobile environment. Thoughtful faculty can apply these principles in the design of mobile learning, leveraging the affordances of technology to create a connected, interactive participatory culture online. The establishment of such a network can promote the creation of an active, collaborative learning system.

A missed opportunity of technology's promise

The unique affordances and potential of new technologies are usually not realised after their initial introduction. Instead, new technologies tend to be used to perform the same tasks as the old technologies that they replaced. It takes extensive familiarity and use for the distinct capabilities of the new technology to be discovered and utilised. The widespread use of *PowerPoint* for classroom lectures is a good example – the computer used as a glorified overhead projector to support business-as-usual lectures (Clark, 2008).

Mobile learning, still immature, has suffered this fate. Professors transform their courses into mobile experiences, assuming that the class will duplicate the on-ground experience. This transition from a bricks and mortar experience attempts to replicate the traditional model of information transfer from teacher to student – attempting to sustain a model that fails to leverage the wealth and opportunities of the evolving digital world. In *A New Culture of Learning* (2011), Thomas and Brown identify the differences between the teaching-based approach to education and the learning-based approach in which digital media provide access to a rich source of information and play, and the processes that occur within these environments are integral to the results. The teaching-based approach focuses on teaching us *about* the world, while the new culture of learning focuses on learning through engagement *within* the world.

The stability of the traditional model of education has been challenged by the very nature of the ever-changing world of technology around it. Information that has long been the domain of formal education has now become accessible to all. Learning information has been replaced with developing strategies for finding information. Static learning, what you learn about a topic, has been replaced by active learning – communicating, experiencing, and doing. Mastery of information in an active world leads to new applications, mash-ups, collaborations, and innovations (Thomas and Brown, 2011).

The current Web 2.0 represents online collaboration and sharing among users (i.e., social networking sites, wikis, communication tools, and folksonomies). This has led to the emergence of the mobile technologies and mobile devices such as smartphones and tablets which are driving the purchasing and implementation of applications within higher education. The future web, often called Web 3.0, is being called "the intelligent Web using semantic webs, microformats, natural language search, data-mining, machine learning, recommendation agents, and artificial intelligence technologies – which emphasise machine-facilitated understanding of information in order to provide a more productive and intuitive user experience" (Spivack, 2014). Gaining access to these databases and webs of the next generation of knowledge communities will be even more rapid as cameras and connectivity continue to improve on mobile devices. Mobile devices will be the primary access point for participatory interaction in which searchers build semantic webs of knowledge assuming that each person has something to contribute to the continued development of knowledge. While the knowledge can be fairly static, the social processes of the Web 2.0 generation for acquiring that knowledge are dynamic – continually testing, reaffirming the work of the collective (Jenkins, 2006).

Sustainability through mentoring

The old hierarchical system of teacher as expert and student as novice worked in a time when the model of higher education focused on teaching and the transfer of information from professor to student. The university had a monopoly on knowledge and the credentials it bestowed were the indicators of academic success. Today mentorship is much flatter, and what an individual knows and can do often trump a formal credential.

Marilu Goodyear (2006) has noted that the concept of the developmental network shifts the focus of mentoring from the top-down approach of organisations assigning mentors to mentees to a more independent approach with the focus on the person mentored as a student as scholar. In this modern concept of mentoring, the responsibility for mentoring sits squarely on the mentee. Mentees develop their own developmental networks in relation to their particular needs. Mentees reach out to individuals around them to seek assistance in the functional areas where they need help.

Mobile devices and access to rich resources and people around the world accentuate the development of tacit knowledge over formal implicit learning that has traditionally been the business of higher education. Reber (1993) and Von Krogh, Ikujiro and Kazuo (2000) differentiate between them: implicit learning is the acquisition of knowledge that takes place largely independently of conscious attempts to learn and largely in the absence of explicit knowledge about what was acquired.

Apprentices in craft guilds in the Middle Ages assumed basic duties like sweeping the floor and fetching materials. By being in proximity to master craftsmen and observing their work, they began to absorb much information about the culture and function of the guild. Over time they were taught to perform simple functions like sewing buttons on garments. Their journey to become a master tailor was conducted over time with expert modeling and informal instruction.

Higher education in the past was considered to be the innovator because there was little competition. In recent years, the ability of higher education to appear nimble has disappeared; the present rapid pace of technology has allowed small and large businesses alike to leapfrog the work of higher education. In the past, scientists gathered to brainstorm, gave a nod to what exists and then thought about what needed to exist, making careful, incremental gains. The research was empirical and built upon past discoveries (Mansfield, 1998).

Today business and industry conduct their own research. It is pioneering and applied. Before a product is developed, ethnographic studies are conducted to understand the consumer; during development prototypes are tested and iterated to create rapid improvements; and customer focus groups are held frequently to understand user reaction and opinion (Anderson, 2009).

These types of activities are often conducted online – not by just the scientists – but by business and youth culture globally. Mobile technologies are a large part of this leapfrogging process in which crowdsourcing, while not the traditional way of introducing products and ideas, has led the way by lending a voice to desired products, medical experimentation and services offered to a community. Rapid protocoling is finding its way into individual and team-led innovation in a way that was not foreseen by higher education. Mobile technologies bring with them a personalisation that has not previously been seen or brought into the higher education classroom (Soares, 2011).

Personalisation is the name of the game for today's learner – one size does not fit all – in today's market as well. Expertise is evidenced by what the learner can do. Mobile devices allow for direct participation in a web-connected world in which mobile learning resources are more convenient than computer-based resources. A learner can make the decision as to when to work on a project, when to seek out others for help and when to opt out or opt in to a discussion that is not permission based. The mobile resources are readily available with the mobile device and can be forwarded to friends and family when needed. The flexibility of availability of the resources coupled with time and place determined by the user situates the learner in charge of not only access but also of the content being distributed to a collective participatory audience (Mobile Learning, 2014). Relationships and work are different for today's digital learner as they use social networks such as *WhatsApp*, *Twitter*, *Facebook*, *Instagram* and *LinkedIn* to communicate with their friends and acquaintances.

Much of Web 2.0 software shares both a real world social world and a virtual world. Software such as *YouTube* and *Instagram* allow for the sharing of photos and videos both in the real world with the friend standing next to you and the virtual world of 'friends'. *Facebook* and others provide a platform for the personalisation of one's profile with video, multimedia, building a community of 'friends' with similar interests. Relationships are established, and before long, the social network seems 'alive' and 'real'; add a mobile device and real-time interactions can exist through the device's microphone and camera (McLoughlin & Lee, 2007) bringing to each a visual and verbal connectivity.

Participatory learning – and in this case, selection of goods and services – is now playing a key role in the innovation of products at a rapid pace. Web 2.0 applications found freely on the web share many of the pedagogical goals sought in higher education: participation, engagement, discussion and collaboration (Grosseck, 2009). These same pedagogical goals while not noted as coming from education are often referenced by business as needed for innovation: collaboration, critical thinking, and discussion skills. Today's innovation and participation has to be seen as conversations that are web dialogues within knowledge communities where people play to learn, reskill and gain a reorientation to the problem to be solved (Grosseck, 2009).

The Open Learning University

Online education is best understood as serving older, non-traditional students who do not have access to a college in their area or cannot attend regularly due to jobs and family. Now the same technologies of distance learning are on bricks-and-mortar campuses, especially public institutions with declining state funds. The implementation and delivery of online education in higher education exceeded 7.1 million students in the fall term of 2012, with an increase of over 411,00 students from 2011 (Pearson, 2014, para. 1). The university campus is more open than in the past due to the online courses now offered to both on-campus residential students and off-campus non-residential students. While the value of online courses continues to be debated, a large majority of students (approximately 3 million) are simultaneously enrolled in face-to-face courses and online courses on their university campus. This belies the popular thought that most online students live far from campuses (Gabriel, 2010).

"At the University of Florida, for example, resident students are earning twelve percent of their credit hours online this semester, a figure expected to grow to twenty-five percent in five years" (Gabriel, 2010, para. 11) due to the simple economics that 1500 students enrolled in a lecture cannot be held in a campus lecture hall. Universities are finding that on-campus students are just as likely to come to campus and attend online courses, bringing to mind the question of the need for the traditional faculty and classroom. While some question the mission of the university in providing face-to-face contact with the faculty and classmates, others note the need to meet a mission of serving as many students as possible while remaining affordable, as well as a desire to exploit the latest technologies (Gabriel, 2010).

Online learning brings into play a different sort of participatory learning. Higher education has increased its delivery of online courses at a rate of 33.5% in the past two years (Allen & Seaman, 2014). This increase has led to a cry of 'any time, anywhere, 24/7' learning opportunities. Allowing access of this nature to an increasingly older population seeking to improve skills, obtaining degrees otherwise inaccessible or finding those seeking new careers, has brought into the university a new group of learners. These learners have a very different expectation for the delivery of their educational experiences. Expectations include a learning that is tailored to their needs for just-in-time learning. Opening up the online world to this group of learners brings some added challenges as they do not necessarily want to be bothered by the instructor but want autonomy to move through material at their own pace and time. Such autonomy is often beyond the reach of traditional students (Allen & Seaman, 2014; Bonk & Khoo, 2014).

The present nature of competition in higher education brings into focus issues of the sustainability of mobile learning. Industry, while actively seeking college graduates, is also rewarding and hiring those who are gaining competency-based educations through the use of Massive Open Online Courses (MOOCs). Groups such as *Coursera* and *Udacity* are building a new expectation within industry that learners can attend, collaborate with others as teams, and learn the needed knowledge, skills and dispositions to meet the workforce needs of society. Coupled with a nod from universities such as Duke University as a validator of the MOOC, learners are working online 24/7 to learn needed skills of content, teamwork, innovation and critical thinking all while participating in and building the course for all involved (Daly, 2013; Ram, 2013).

"MOOCs are forcing institutions to learn how to collaborate, share resources, and shake off an historical institutional-centric world view" (Galer, 2014). MOOCs provide valuable information about how students learn. "We can get hundreds of thousands of data points from students to look at the best ways to present information, which exercises are really productive, and which do students look at and learn from the most", says Jablokow (cited in Galer, 2014). These data points can help to further tailor to the mobile environment of the learner. The faculty member has to be the guide on the side as the collective intelligence of the MOOC members begins to build live participatory networks of learners developing teamwork skills, critical thinking and problem-solving with experts both inside and outside the MOOC.

Sustaining mobile learning in higher education

"We need digital models of learning and teaching. Not just a technology overlay on old modes of teaching and learning" (Professor Venkat Venkatraman, cited in Brynjolfsson & MacAfee, 2014, p. 212). For mobile learning to be successful in higher education, course designers must be mindful of the rapid development and proliferation of new, networked technologies and universal user access to a limitless resource bank of people and materials. While it is impossible to know about and be proficient in every emerging technology, it is important to participate in a network of early adopters and users to stay abreast of developments. A survey of the best technology in education as well as best of class uses in other domains like business and industry will provide insights and techniques for sharing ideas, connecting, forming affinity groups, and posting products. Interviews or focus groups with students who comprise the target audience for mobile learning reveal their preferences and capacities for technology. Rather than attempting to replicate the traditional model of top-down information transfer, mobile learning needs to sustain a model that is nimble and quick to incorporate the unique features of new technologies, leveraging the social nature of networks and the limitless resources that have been made available to all learners.

Threshold conditions

A set of *threshold conditions* creates a framework of minimum requirements that need to be present as precursors of mobile learning sustainability. These common sense threshold conditions offer the capacity to allow new forms of digital learning to flourish.

1 *Responding to context* – Context provides a frame for learning and action as part of reaching a sustainability model. It can be confining or freeing based on its components. At its best, context provides grounding and stability while at the same time offering flexibility to allow for change and evolution. Context is an intrinsic part of the way learning occurs in the real world, in real life – both in how learning takes place as well as the problems encountered in it. Most learning that takes place throughout a person's life is learning that is situated in the real world and in real life. To a large extent, the context for learning provided by higher education has been abstracted away from real world experiences, and the pressure of education to come up with the 'right' answer has made institutional learning stressful rather than exciting (Fitch Richardson Smith, 1990).

2 *The ability to form teams and gain access* – Sustainability involves the use of robust bandwidth to guarantee access to the use of online tools and mobile resources. This bandwidth must be available in rural,

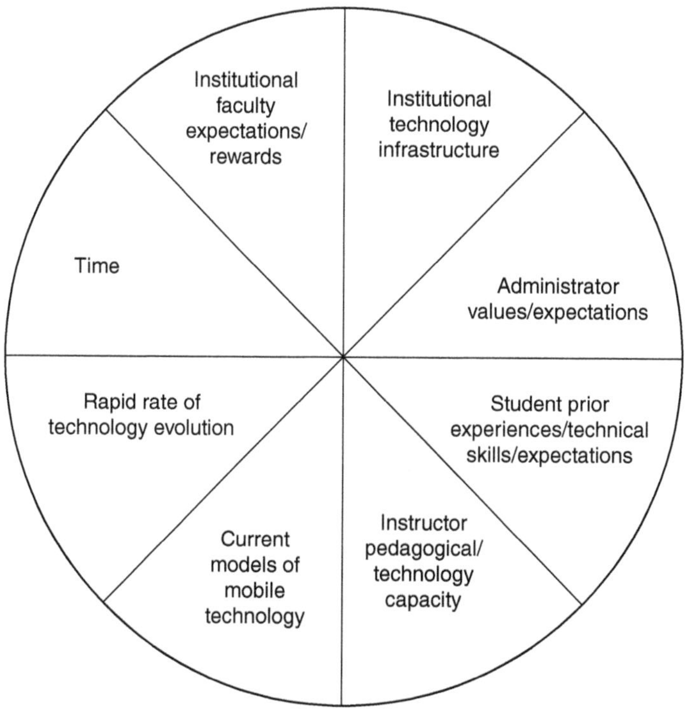

Figure 2.2 Unfreezing the current meta-stable institutional system (context for mobile learning) (These factors interact and impact one another to inhibit change and innovation. Policy changes, new institutional priorities and needs, and risk taking by teacher/leaders are some of the things that can support institutional change.)

urban and suburban communities at a reasonable cost such that 24/7 mobile learning is a reality. The importance of active participation and the ability to leverage the development of skills for collaboration is critical in a networked world. Content is not enough; it is important to teach and model forming and participating in teams that can challenge known content. "None of us know everything; each of us knows something; and we can put the pieces together if we pool our resources and combine our skills" (Jenkins, 2006). The ability to pool our resources sets in motion a convergence of power. Collective meaning making can change how institutions of higher education operate to provide greater connections to the world for authentic learning (Jenkins, 2006).

3 *Design* – Sustainability realises the importance of learning discernment regarding accuracy and quality of web resources and leveraging experts and practitioners to gain multiple perspectives. It is equally important to understand the intended audiences' needs and latent needs – needs they don't even know they have – as well as the authenticity and accuracy of resources. Business and industry design with one eye on the consumer and the other eye on their competition. Mobile learning gives higher education the opportunity to reconceptualise the curriculum design process. To better understand the users, conduct ethnographic studies to understand potential learners; discover how do they learn tacitly and formally on their own, what technologies do they use for personal learning, and what do they seek and value regarding personal learning. Find and study institutions and businesses that are successful early adopters of mobile learning – the best of brand. Analyse their success and adopt their best practices in your design. Survey peer institutions to understand how they are addressing mobile learning at their institutions. Learn to package your ideas using a range of media to inform and engage your audience and seek frequent feedback to assess your success and tweak your approaches as needed.

Grades are more important than learning. Value is placed on what we can grade (the measurable outcomes of academic activity) – rather than learning to grade what we should value (conceptual understanding, innovation and creativity, and discernment). "Grading has become the core function of the modern system of higher education. Nearly every process in the university relies on grades from matriculation, class enrollment, financial aid, and graduation, to departmental accreditation and a school's reputation. In addition, the heavy emphasis on grades by institutions of higher education and private and public sector employers has caused students to fetishize the grade to the detriment of learning" (Canally, 2012). The current reward system is so deeply entrenched that it seems next to impossible to create a feedback loop that focuses on ephemeral aspects of learning and understanding. A new model of rewards that prizes doing as well as knowing will legitimatise the enterprise of formal learning.

4 *Incentives* – Sustainability requires incentives for faculty that reward high-quality teaching and student learning in which the faculty member becomes a facilitator allowing for participatory engagement of not only the learners enrolled in the classroom but the collective learners outside the classroom. Opening the classroom to a wider variety of participants to challenge the learner creates a more innovative discussion and problem-solving culture of learning that includes novices as well as experts.

5 *Legacy* – Sustainability in higher education must begin with an understanding of the affordances of technology that allow for the development of a legacy knowledge in which learners are able to build complex systems through partnerships with organisations outside the institution and living on the Web. This legacy knowledge becomes the foundation for interdisciplinary work to solve the complex problems facing mankind building a reservoir of ideas, innovations and entrepreneurship to power new learning.

"Sustaining the unsustainable: reframing global learning for the twenty-first century" acknowledges a fundamental challenge of our times: how unprecedented technical capability in the form of mobile learning can be used to meet current vocational and economic challenges. This is particularly true in a world where routine work once done by human labour is increasingly replaced by more cost-effective technologies that can be programmed to complete such tasks (Brynjolfsson & McAfee, 2014). Accepting the premise that this phenomenon is likely to accelerate in the future highlights the need for pedagogic and andragogical changes described above. Learners must be prepared to compete in a world where capital expenditures on technology increasingly replace what has formerly been spent on human labour. Those with skills that maximise the labour-saving affordances of technology or who perform work that isn't capable being programmed are likely to be far better prepared to compete. The former group includes people that will be able to recursively analyse data, frame questions and draw inferences or conclusions based on creative higher order thinking skills. The latter includes those engaged in vocations that do not deliver routine services, such as machine repair, design services, landscaping, and numerous health care enterprises (Brynjolfsson & McAfee, 2014).

To meet this challenge, education must acknowledge a difficult lesson that business and industry have already learned about technology – that for every dollar spent on new hardware as much as another nine dollars must be spent on software, business process re-design and professional development (Brynjolfsson & McAfee, 2014). The winners in this increasingly competitive world are likely to be those who effectively confront the realisation that what is not sustainable is a model of education that uses mobile learning adaptively or believes the mere possession of mobile technology devices will bring the transformation that meets the challenges that are already upon us.

Facilitation in higher education as a best practice

Finally, the higher education community must acknowledge that teaching and learning is not an innate skill or endeavour. The ability to facilitate student learning is a *learned* endeavour, and faculty around the world must be prepared to facilitate learning to a mobile community of learners that can be connected to a large group of experts. Higher education administration and faculty have often opined about moving from faculty-centred learning to student-centred learning without much success, as most courses on university campuses are still dominated by lectures and text-based learning in a face-to-face classroom – even the online courses of many universities are mostly comprised of lectures.

Preparation of higher education faculty to facilitate learning has to be intertwined with an understanding that curriculum in a discipline now needs the integration of technology, leveraging the unique affordances of technology to provide learning resources and experiences that could not happen otherwise. The nature of the continuous, rapid change of technology makes the practice of recycling lectures and curriculum ad nauseam obsolete. These constant changes force continuous reinvention and iteration of the curriculum to keep the material current and appeal to the ways that student learning evolves. With that change comes a different approach to how learners engage with the content. Stop thinking only pedagogy and think engagement and mobility. Encourage faculty to expand their repertory of learning tools to include pedagogy, andragogy, and a model that encourages critical thinking and problem solving.

Today's higher education students are mobile and connected and are often spending more for smartphone access than for a textbook. The higher education community has noted the need for ubiquitous access on campus by providing laptops to checkout and offering unlimited bandwidth in dorms and campus coffee houses. The need for courses that are flexibly structured to maximise the use of time and accommodate the varied lives of their students, staff, and community is critical to this new generation of learners who often work full-time and go to school part-time to avoid debt.

Mobile technologies support the flexibility of this academic structure by allowing learners to stay connected and respond as their schedules allow. Assignments continue as the learner moves through a variety of environments – always connected to his/her personal learning device and with team members to complete assignments. The structure of higher education programs must extend beyond the traditional hourly course day and semester-based year as well as beyond the campus. The next generation of higher education programs must draw on the entire community's ever-expanding inventory of global resources to foster a team for learning that is authentic and is situated in a real world context (SERC, 2014).

Facilitation by faculty to this approach to learning must be continually adjusted on the basis of data collected through a variety of valid and reliable methods that indicate student progress and needs. The assessment results can be interpreted and applied appropriately to improve individual student performance aligned to support rigorous and challenging learning that is in a real world context and uses mentors and "virtual" experts to support learning (SERC, 2014). Learners should be challenged to be actively involved in their learning through inquiry, in-depth learning, and performance assessments not the typical multiple-choice test (McGee et al., 2005). The movement to more performance-based assessment also allows learners to take a realistic measure of their own learning through the design of a product thus ending the need for "grading" that is unrealistic and does not represent what occurs in the real world of product design and development. Grading by design, development and performance creates a network of products connected by a wide variety of internal student/learner teams with external experts continuing the legacy of knowledge development to foster innovation within higher education classrooms.

Moving to sustainability

Rather than attempting to replicate the traditional model of top-down information transfer, mobile learning needs to be supported by a model that is nimble and quick to incorporate the unique features of new technologies, leveraging the social nature of networks and the limitless resources that have been made available to all learners. The irony of sustaining mobile learning – which itself is something of a moving target – is that we need a dynamic, protean platform that can adjust quickly and react to new technologies and their social application to provide ongoing support. We need to reconceptualise the enterprise of higher education to enable it be relevant and valuable in a time of rapid global change.

This chapter explored this new model that sustains continuous change in a new culture that focuses on learning. Much like the dramatic transformations that have been caused by networked technologies in business and industry, higher education will need a significant recalibration to reinvent and energise the existing system and culture. Change is difficult for a closed system such as higher education because by its very nature its mission has been to maintain the status quo and perpetuate its legacy. Moving a rigid meta-stable education system by focusing on a few critical components is almost impossible. What is called for is a reinvention that disrupts the conditions that support the status quo including the role of the teacher; the linear, time-based class; the heavy dependence on lecture and text; the emphasis on discrete, bounded factoids; the focus on prescribed content and individual learning; and the need for new assessments that demonstrate and apply conceptual mastery. In addition, context sensitivity will be critical for success: factors including designing mobile learning to align with the range of technical skills of students; using a repertoire of pedagogies to reach all students; providing abundant bandwidth and technical support; and teachers with strong process skills of facilitation and collaboration. Institutional rewards and supports will be critical to incentivise faculty to embrace a learning model where learning is open-ended and unpredictable; where they are encouraged to explore and leverage unique affordances of technology, adapting and evolving on the fly; and where they can engage students as collaborators – and even as mentors when students contribute expertise to the learning enterprise.

The model proposes structures that provoke exploration and invention rather than prescribe a discrete, bounded process. This emphasis shifts from knowing about something to applying new knowledge in context to create something. The result treats knowledge as fluid and evolving and is well aligned with the relentless change that has become the new world order. This student-centred model moves beyond transferring information to allow learning to become transformative and alive, engaging and exciting the learner in ways that lectures and books never could.

References

Allen, E., & Seaman, J. (2014, January). *Grade change: Tracking online education in the United States*. Babson Survey Research Group and Quahog Research Group, LLC. Retrieved May 3, 2014, from http://www.onlinelearningsurvey.com/

Anderson, Ken (2009, March). *Ethnographic research: a key to strategy*. HBR.org. Retrieved December 7, 2014 from http://hbr.org/2009/03/ethnographic-research-a-key-to-strategy/ar/1

Bonk, C., & Khoo, E. (2014). *Adding some TEC-VARIETY: 100+ activities for motivating and retaining learners*. Retrieved April 10, 2014, from http://tec-variety.com/

Bransford, J. D., Brown, A., & Cocking, R. (2000). *How people learn: Mind, brain, experience, and school*. Washington, DC: National Research Council.

Brynjolfsson, E., & McAfee, A. (2014). *The second machine age: Work, progress and prosperity in a time of brilliant technologies*. New York: W.W. Norton & Company, Inc.

Canally, C. (2012). *Intervention—where's our agency? The role of grading in the neoliberalization of public universities*. Antipode Foundation.org.

Chickering, A. W., & Gamson, Z. F. (1987). Seven principles for good practice in undergraduate education. *AAHE Bulletin*, 3–7.

Christensen, C.M., Johnson, C. W., & Horn, M.B. (2010). *Disrupting class expanded edition: How disruptive innovation will change the way the world learns*. New York: McGraw-Hill.

Chubb, J. (2014, April 9). *Transforming via technology: Competition and choice*. Retrieved May 1, 2014, from http://educationnext.org/transforming-via-technology-competition-choice/

Clark, J. (2008). PowerPoint and pedagogy; maintaining student interest in university lectures. *College Teaching*, *56*(1), 39–44.

Daly, J. (2013, June 25). *Randy Riddle, a Duke technology consultant, explains how and why the university uses MOOCs*. Retrieved April 10, 2014, from http://www.edtechmagazine.com/higher/article/2013/06/inside-look-duke-universitys-mooc-initiative

Fitch Richardson Smith (1990). Learning and technology: new perspectives. Research conducted for Apple Classrooms of Tomorrow, Apple Computer.

Franklin, T. (2010). Mobile learning: At the tipping point. *Turkish Online Journal of Educational Technology – TOJET, 10*(4), 261–275.

Gabriel, T. (2010, November 4). Learning in dorm, because class is on the web. *The New York Times*, Retrieved April 15, 2014, from http://www.nytimes.com/2010/11/05/us/05college.html?pagewanted=all&_r=0

Gabriel, T. (2010, November 5). Live vs. distance earning: Measuring the differences. *The New York Times*, Retrieved May 5, 2014, from http://www.nytimes.com/2010/11/05/us/05collegeside.html?ref=us

Galer, S. (2014, May 12). *How MOOCS are disrupting education.* Retrieved May 12, 2014, from http://scn.sap.com/community/business-trends/blog/2014/05/12/how-moocs-are-disrupting-education

Goodyear, M. (2006). Mentoring: A learning collaboration. *Educause Quarterly, 4*, 51–53.

Grosseck, G. (2009). To use or not to use Web 2.0 in higher education? *Procedia Social and Behavioral Sciences, 1*(1), 478–482.

Hacker, D. J., & Niederhauser, D. S. (2000). Promoting deep and durable learning in the online classroom. *New Directions for Teaching & Learning. Winter 2000* (84), 53–60.

Jenkins, H. (2006). *Convergence culture, where old and new media collide.* New York: New York University Press.

Jenkins, H., & Deuze, M. (2008). Convergence culture. [Editorial]. *Convergence: The International Journal of Research into New Media Technologies.* Los Angeles: Sage. DOI: 10.1177/1354856507084415

Klein, Daniel B. (2005). The Ph.D. circle in academic economics. *Econ Journal Watch*, 2(1), 133–148.

Mansfield, Edwin (1998). Academic research and industrial innovation: an update of empirical findings. *Research Policy, 26*, 773–776.

Matusky, R. (2014). *Web 2.0 vs. Web 3.0 – What really is the difference?* Retrieved October 8, 2014, from http://diaryofanelearner.com/2013/04/10/web-2-0-vs-web-3-0-what-really-is-the-difference/

McGee, P., Bothra, J., Gurrie, J., & Jadav, A. (2005). If we knew then what we know now. In G. Richards (Ed.), *Proceedings of World Conference on E-Learning in Corporate, Government, Healthcare, and Higher Education 2005*, pp. 3035–3038. Chesapeake, VA: AACE.

McLoughlin, C., & Lee, M.J.W. (2007). Social software and participatory learning: Pedagogical choices with technology affordances in the Web 2.0 era. *Proceedings Ascilite Singapore 2007*, 664–675.

Mobile Learning. (2014). *10 reasons why mobile learning matters.* Retrieved January 15, 2014, from http://mlearning.edublogs.org/2007/01/16/10-reasons-why-mobile-learning-matters/

Pearson. (2014, January 15). *Babson study: Over 7.1 million higher ed students learning online.* Retrieved on December 7, 2014 from http://www.pearsoned.com/babson-study-over-7-1-million-higher-ed-students-learning-online/#.U0s7e8exS8w

Ram, P. (2013). From one-to-one to many-to-many: Power peer learning in open learning environments. *eLearn*, Volume 2013 Issue 10, Article No. 2, 12–17.

Reber, A. (1993). *Implicit learning and tacit knowledge: An essay on the cognitive unconscious.* New York: Oxford University Press.

SERC. (2014). *Evolution of the higher education faculty role: From facilitator to orchestrator*. May 5, 2014, Retrieved on December 7, 2014 from http://ctserc.org/s/index.php?option=com_content&view=section&id=8&Itemid=28

Soares, Louis. (2011). The personalization of higher education; using technology to enhance the college experience. *Center for American Progress*, Retrieved on December 7, 2014 from http://www.americanprogress.org/issues/labor/report/2011/10/04/10484/the-personalization-of-higher-education/

Solar, M., Sabattin, J., & Parada, V. (2013). A maturity model for assessing the use of ICT in school education. *Journal of Educational Technology & Society*, *16*(1), 206–218.

Spivack, N. (2014). Web 3.0. *The third generation of web is coming*. Retrieved October 8, 2014, from http://lifeboat.com/ex/web.3.0

Swail, W. S. (2002). Higher education and the new demographics: Questions for policy. *Change*, *34*(4), 14–23.

Thomas, D., & Brown, J. S. (2011). *A new culture of learning: Cultivating the imagination for a world of constant change*. [Self-published].

Von Krogh, G., Ikujiro, N., & Kazuo, I. (2000). *Enabling knowledge creation: How to unlock the mystery of tacit knowledge and release the power of innovation*. New York: Oxford University Press, p. 7.

3 Waypoints along learning journeys in a mobile world

Agnes Kukulska-Hulme and Mike Sharples

Introduction

The metaphor of a 'learning journey' evokes an abstract idea of progress or development that may also be facilitated by physical movement or travel across a terrain. This metaphor has currency in our increasingly mobile world, where travel is common and ubiquitous technologies may shape the journey, preserve it in multiple media and enable it to be widely shared. One quandary when considering a learning journey is that it may appear to be a simple linear progression from A to B, while at the same time it is a component of a larger and more complex journey that may even be equated with a life. While mobile learning approaches aspire to make learning "seamless" (Chan et al., 2006; Wong and Looi, 2011), we note that specific episodes of learning with their associated travel or wayfaring still need to be recognised so that they can be investigated in research or shaped for specific educational ends. Also, effective learning involves metacognition (Bransford, Brown, & Cocking, 2000) as a process of extracting meaningful episodes to reflect upon from the flow of experience.

Learning journeys evoke some familiar patterns, but there is no universal concept. A learner's journey usually includes significant points when goals are set, progress is tracked, and outcomes may be noticed, assessed or shared. In literature, education and in life, learning journeys are commonly described in terms of personal accomplishment which involves facing some challenges along the way. However, perspectives vary across different cultural settings. Chinese students learning English have been observed to portray their learning as "a bitter-sweet journey of ups and downs towards ideals and dreams of future success" (Jin & Cortazzi, 2011, p. 113), while Marie Martin, in the Irish tradition of 'learning by wandering' (exploring with an open mind, without schedules or planning), depicts the learning journey as one in which "the process is its own reward" (Martin, 2010, p. 24). Learning, she submits, is a life-long commitment: "There is no end point. There is always more to learn" (*op cit.*, p. 24). In his monumental study of factors affecting learning outcomes, Hattie identifies the essential role of a teacher or mentor in offering a succession of appropriate challenges and ensuring the learner is on the right path to meet, understand and benefit from them (Hattie, 2009, p. 38).

Our contemporary societies are grappling with the issue of how to sustain life-long learning and continuous professional development in view of economic imperatives and trends toward more frequent job and career changes over a lifetime. Current educational advancement is typically determined by life stages and professional requirements, amounting to educational episodes that have definite end points based on recognised achievements. To sustain lifelong learning, there is an emerging consensus that more diverse opportunities to learn and study should be offered that provide quicker routes to completion, "flexible schemes to accumulate credits before, during and after a traditional programme of study" (Gordon, 2013, p.3), and more varied ways of reporting on progress and certifying achievement. All this implies that learning routes and pathways need to be considered afresh, and the contributions of learning journeys need to be explored.

Learning for life may be conceived as a continuous process involving exploration, skill development through practice, conversation with peers, and reflection on experience. From a professional and career development point of view, it frequently needs to be focused, applied and perhaps evaluated as to whether it helped bring about a desirable change. Projects are one connection between the two conceptions of learning – one more open-ended and the other more focused – where the learning projects may be personally or formally initiated, and personally or formally supported. Learning projects have the characteristic of being extended over time and locations, hence leading to learning journeys that may be well supported by ubiquitous technologies helping to organise a sustained sequence of learning episodes that leverage social networks, time and space.

In this chapter we consider how mobile and ubiquitous technologies support learning journeys, with particular reference to significant landmarks along these journeys that may help sustain learning in the longer term. We set these considerations in the context of learning journeys past and present.

Learning journeys in the past

The notion of learning as a journey toward enlightenment and truth goes back at least to Confucius. It was adopted by humanitarian educators including Montessori, Steiner and Dewey and has been explored more recently through holistic education, where learning is seen as a quest to find identity, meaning and purpose (Cajete, 1994; Miller, 1996; Palmer, 2004). In this sense, a learning journey is an integrative process whereby a love of learning and a sense of wonder lead people to engage with ideas and environments guided by a supportive community. The formal education system of curricula, attainment targets, key stages and examinations presents a different type of learning journey in which the goals and stages are set by the institution, and the learner is required to navigate through these to attain qualifications. Lesson planners for teachers and e-portfolios for learners are tools to create and record these formalised pathways.

Linking journeys for holistic and formal education is the concept of narrative, a personal and collective story. So, individuals create life stories which they chronicle as memories, notes and images to make meaning from experience. These personal narratives link together into communities of practice (Lave & Wenger, 1991) where the experience of learning is formed through shared activity and common goals. They are situated in private and public spaces where people come together to discuss ideas and engage in mutually supportive practices. The school classroom is one such space: students are required to meet daily to access and create resources for their formal learning journeys. Non-formal spaces where learning journeys intersect include museums and heritage sites, and more liminal spaces such as town centres, festivals and sports events. How to create technology-enhanced 'micro-sites for learning', for example on field trips or in workplace meeting areas, is an area of current research in technology-enhanced learning (Vavoula & Sharples, 2009).

The romantic notion of learners in mutually supportive communities taking shared journeys sustained through a love of learning is at odds with the everyday lives of many schoolchildren – stuck in a classroom, disengaged from the curriculum, or revising for exams. A central issue for educational researchers to address is whether formal and informal pathways of learning can and should be bridged, and if so how. Should children be encouraged to bring their personal learning projects into classrooms? Should schools attempt to structure the learning of children outside the classroom? And should technology provide a bridge between holistic and formal education? Researchers working with adult learners including university and college students face similar questions. Connecting formal and informal learning is an elusive but important educational goal.

Another notion of learning journeys comes from research into learning trails – movements through physical or online spaces that offer opportunities to gather knowledge, engage in conversations, and interact with the world (Reynolds et al., 2008; Sharples et al., 2013). Juliet Sprake (Sprake, 2011) has investigated learning-through-touring: how a tour is an evolving conversation between guide, visitors and the environment. Tour-enabled buildings, such as museums or heritage sites, impose limits on the journey, restricting the making of meaning to what can be seen front-of-stage in the visitors' areas. The un-toured spaces, that are off limits for visitors, may show the museum as a work in progress, where meaning is made by curators with more uncertainty than is shown to the public. Thus, a tour is a line of tension between a displayed environment and an unfolding story of its fabrication.

The learning trail may be formalised in a museum visit or school field trip, or it may be serendipitous, as when visitors to a city stumble upon interesting sites and unexpected events. Sprake (2011) proposes three attributes of a tourist as active learner: stumbling upon, noticing, and connecting. By stumbling upon spaces in unexpected ways visitors create narratives of their experiences. They notice some parts of the environment in ways that are signed and proposed, and others in ways that create personal meaning, so that touring is an act of imaginative association between the given and the found. As they view the environment from different perspectives, visitors enter into conversations and juxtapositions that create cognitive and social connections. In that sense, learning trails create what Dewey called 'learning through occupation' where sensation and thought are combined through embodied activity (DeFalco, 2010).

On a larger scale, journeys through physical and conceptual spaces become life projects. Equipping people to be resourceful and self-determined learners in their journeys through life is a goal not only of formal education, but of successful parenting and effective professional development. This is where learning trails and holistic education meet in the notion of learning as a sustained journey towards a self-determined goal. Hattie (2009) indicates the importance of continued commitment to knowing where to go next, in relation to the gap between a learner's current knowledge and the criteria for success.

Scientists and engineers have traditionally written journals to record their learning journeys, while artists or designers compile portfolios to show their creative journey as well as showcasing their work. In formal settings, the journal can be a tool to increase student engagement and reflection on learning. For example, Park (2003) found many positive outcomes for learning journal use in geography, including increased student interest in and engagement with course material. Personal journals have long had a role in promoting personal growth and learning. In an exploration of the uses and benefits of journal writing across various spheres of life, Hiemstra (2001) concludes that they can encourage a proactive approach to the learning process. We will revisit the role of journals in our later section on 'Capturing progress and sharing achievement'. In the next section we consider how learning journeys are changing and how they might further evolve.

New learning journeys

Following in someone else's footsteps or following directions set by a teacher or a guide becomes a different experience when the learner is a more active participant and when recordings of what happened on the journey may be easily made and shared. There are various examples of how mobile technology has been used to facilitate an active experience. Museum and art gallery tours, as well as trails in natural environments or facilities such as botanical gardens, have been fertile ground for innovation in this regard. In an early research project, schoolchildren played a highly active role in Mudlarking in Deptford (Sutch, 2005) by co-producing a guided tour of a riverbed in London, the Deptford Creek. This is an area of historical interest, and it is also of ecological interest since it is home to hundreds of fresh and saltwater animals and plants. In this project, a handheld device with GPS capabilities delivered location-sensitive information when a child walked into node areas indicated on a map; the children could also create their own multimedia content and alert other users to that content. In the MyArtSpace project (Sharples et al., 2007), an explicit connection between active learning in a museum or art gallery and reflection and sharing in the classroom was made through a combination of mobile technology and a website accessed on a school desktop computer. Schoolchildren used mobile phones during a visit to a museum or arts centre to access pre-prepared information in context, take photos, make voice recordings, write notes and see who else had viewed the same exhibits. The content was transmitted to a website which stored a personal record of each child's visit, giving them access to their collected materials which were then used to create presentations back in the classroom.

Both the above projects are examples of reconceptualisations of the trail or guided tour, enabling participants to add their personal experiences for later reflection and reworking, or benefiting of future tour users. Both projects gave students opportunities to create memorable journeys and to collaborate in novel ways. An outdoor activity can also be a helpful way to track learning in terms of conceptual change. To support changes in concept formation, an interactive concept map approach for supporting outdoor mobile learning activities was proposed and evaluated by Hwang, Wu and Ke (2011). The elementary school students were asked to create concept maps about butterfly ecology, then to observe the ecology in a real-world environment and revise their concept maps if necessary. The PDAs used by the students assisted them in the field by evaluating their concept maps and giving instant feedback or learning guidance.

In the above example (Hwang, Wu, & Ke, *op cit.*), a note-taking function on the PDA also allowed the students to describe what they had found or to raise questions based on their observations. Such note-taking activity could potentially lead to further research or observations. Song, Wong and Looi (2012) mention that their mobile science inquiry activities with school children explicitly evolved to give the learners more scope to determine the direction of their learning: "Though starting with a relatively structured mobile learning trail in the farm, we subsequently facilitated and even encouraged the students to show their diversity in deciding what to learn and what artifacts to produce" (p. 698). Learning outside the classroom gives opportunities for learners to conduct their own contextual inquiries and perhaps to develop more autonomy. However, this cannot be taken for granted. In the field of mobile, situated language learning, the experiences reported by Abbas Petersen, Procter-Legg and Cacchione (2014) of young adult learners using a mobile app designed to help them capture and share language elements they come across in their everyday lives, suggest that learners may not be ready for autonomous learning. The authors conclude: "We need... to teach them not to be taught, i.e. to become autonomous learners" (p.70). Tan, So and Zhang's (2012) research study investigated students' autonomy in an inquiry-based mobile learning trail, taking into account the design of the trail, interaction with teachers and collaboration with peers. They report that students' capacity for more autonomy in mobile learning was contingent on student readiness, learning design, technological mediation, and the community of learners. This suggests that while mobile inquiry-based learning may be conducive to developing autonomy, other factors also come into play. A learning journey is partly a personal experience, reminding us of the individual's relative capacity to act in self-determined ways, and partly a collective one, drawing on various resources including other learners.

The learning journey continues to fire the imaginations of educators, as cross-disciplinary thinking and technological developments foreground social interaction. The notion of co-learning has been gaining in popularity as new technologies can support collaboration in spite of physical distance and among more diverse groups of people. Thus in Canada, an integrative science university program brings together indigenous and mainstream sciences and ways of knowing through a "co-learning journey" between recognised holders of traditional ecological knowledge and others with expertise in mainstream science (Bartlett, Marshall, & Marshall, 2012). Co-learning denotes an approach in which learners, educators and other stakeholders participate jointly in the learning process; Hannon, Collins and Smith (2005) applied this to education in graduate entrepreneurship, for instance, while Worswick et al. (2014) have used co-learning in an interprofessional quality improvement initiative in primary health care services. Worswick et al. describe what happens when individuals embark on a learning journey together:

> By sharing their experiences and their stories they can come to share beliefs, behaviours, perceptions, understanding and goals for what they want to achieve through the project. They progress towards making and influencing changes through their shared learning experience. By sharing their stories, a new story is created and previously held beliefs might change. With a renewed understanding, learners can achieve their shared goals within the co-learning relationship. Learners can become empowered to lead the change.
>
> (Worswick et al., 2014: 3)

A co-learning journey may be facilitated through the use of social and mobile technologies that can be used across diverse settings on a continuous basis, but in other respects the success of the technology-supported co-learning journey is not guaranteed. In the above example, a physical environment was created where learners felt equal, safe and secure about being open in their discussions and disclosures, and it was suggested that this combination created opportunities for the co-learning partnership to evolve. Technological environments open up opportunities for equal collaboration, but safety and security concerns are more difficult to resolve. A related point, that “people learn from trusted networks”, is made by Tsui, Tsui and See-To (2013) who have been working on a Personal Learning Environment and Network to support peer-based life-long learning; in these personal environments, learners can select which tools best fit their learning purposes and they can use them “to build up networks for co-learning and locating expertise” (p.49).

Co-learning journeys are an implicit concept underpinning the MASELTOV project (Gaved et al., 2013; Kukulska-Hulme et al., 2012) which introduces a novel approach to supporting informal learning journeys undertaken by immigrants settling down to life in a new city. Since social integration is a two-way process, natives and immigrants potentially have a great deal to learn from one another. The MASELTOV project is developing a suite of smartphone-based services and tools to enable users to explore the city and use it in a variety of ways as an informal learning environment. Along with tools for navigating, translating signs, and so on, there are social tools that enable learners to get in touch with people nearby who might help them. These people could be registered members of a community or local vetted volunteers. The technology opens up many avenues for informal, incidental learning underpinned by a recommender system that can use location tracking to suggest relevant learning resources and places to visit. At the same time, there are challenges around providing a safe and secure learning environment to a target group that may be vulnerable or fearful about their location being tracked. Although tracking is not mandatory (it can be switched off by the user), the project highlights issues surrounding potential benefits and challenges of new open learning environments such as the city.

Issues of trust and privacy pervade many contemporary discussions of emerging technologies that enable tracking and recording of learner movements and location-based interactions, as well as those that support experience capture, including wearable devices like *Google Glass* (Hong, 2013). The ‘life-logging’ technology employed in the Ubiquitous Learning Project reported by Ogata et al. (2014) aims “to capture learning experiences in daily life and reuse them for learning and education” (p.98). Life-logging can be done actively by the learner, but in passive mode a camera is worn and takes pictures automatically which can later be retrieved, reflected upon and analysed. Life-logging encompasses collaborative learning in that the captured ‘Ubiquitous Learning Log Objects’ can also be shared.

For the future, contextual technologies could enable deeper engagement with the environment, by offering facilities to make location-based notes, to annotate objects, buildings and landscapes, and to leave virtual graffiti and stories for others to discover. As many people’s learning trails intersect in time and space, there are opportunities to mark the points of crossing, through pictures and recorded conversations. This is already starting to happen in the online environment, through sites such as *TripAdvisor* where people can leave their impressions of a hotel, restaurant or location to influence the journeys of others. Writing a review also provides an opportunity for the writer to create a waypoint and reflect on the journey so far. Ubiquitous technology will mean that learning journeys can now be supported and reviewed almost anywhere: in the countryside and in cities, across campuses, workplaces, heritage institutions and places for organised leisure. Collaborative learning and co-learning, facilitated by learner creation of content and content sharing, can all enhance the journey. The scope of learning is extending beyond the classroom to pathways through real and virtual worlds. But how do learners know that they are making progress and how might their progress and achievement be recognised? A challenge for schools and other institutions is to engage learners with journeys for learning, towards valued goals.

Capturing progress and sharing achievement

If we see learning through the metaphor of a journey over time, across locations, with others, then significant waypoints on that voyage can be recorded and recognised. At the most personal level, experiences of learning are captured through a person's episodic memory, where they remember elements of personal significance from the flow of everyday experience. The familiar technologies of notebooks, diaries, cameras, and audio recorders can help preserve those autobiographical events. Tulving (1985) refers to "autonoetic consciousness" that is a personal ability to appreciate one's personal past and future – what Wheeler, Stuss and Tulving (1997) refer to as "mental time travel". Autobiographical aids such as notebooks and cameras assist that travel backward and forward in personal time not only by recording an event at a particular time and thus connecting it to an explicit timeline, but also by acting as an index back to the personal experience of the event, thereby enriching one's memory by restoring aspects of the context in which it was first experienced.

Gemmell and colleagues developed MyLifeBits as a "surrogate memory" to store everything that could be recorded of a person's online information including articles, photos, emails, web pages, phone calls and room conversations (Gemmell, Bell, & Lueder, 2006). These can be accessed through a timeline of events or a map of trips. In more recent years, the personal data that can be stored range from heart rate and EEG (brain activity) traces, to personal events captured on CCTV security cameras. The problem that arises is how to make personal meaning from this undifferentiated mass of recorded personal experience. Vavoula and Sharples (2009) discuss "lifelong learning organisers" that enable a person to organise everyday events into activities, episodes and projects, representing increasing levels of sustained experience. Vavoula and Sharples' KLeOS system (Vavoula & Sharples, 2002) shows these on a timeline as lines and points, and also allows the user to link them to notes in a concept map, as a way to connect external depictions of a person's episodic and semantic memory. The aim of KLeOS was to enable a person to record a sustained flow of events, while creating order and meaning from the experience.

As more people capture experience, these trails and timelines interweave. Not only can the life tracks of many people be shown and shared through *Facebook*, *Twitter* and other social media, but they can also become resources for learning. Here, we should make a distinction between a shareable record of learning and a record of shared learning. A record of learning, such as an e-portfolio, which records a person's evidence of learning, can be shared with a tutor or other learners. For example, the UK National Health Service provides an e-portfolio to more than 300,000 health service professionals to record learning activities and assessments which can be kept private or shared with a supervisor (www.nhseportfolios.org). The *LinkedIn* site (www.linkedin.com) allows users to record and share their significant learning projects and achievements. Although its premium site offers a *Twitter*-like feed of activities, it does not yet offer the facility to log the detail of learning episodes (for example, to indicate books being read or courses currently being taken).

A record of shared learning is a social space to collectively record and share learning experiences. An example is the *iSpot* site (www.ispotnature.org). This allows any registered user to observe, photograph, identify and share observations of wildlife. Others in the community confirm the identification, building up a shared record of observations of the natural world which can then be indexed by time, location or species. Like KLeOS, the site also allows observers to record meaningful links between observations to create a concept map (for example, by recording a 'visiting' link between an insect being observed and the flower that the insect was visiting).

The *iSpot* site also supports reputation management and a badging system (Clow & Makriyannis, 2011) which is another piece in the jigsaw of capturing progress and sharing achievement. The central purpose of a reputation system, whether it is automated as in *iSpot* or manual as in Scout and Guide badges, is to recognise and reward legitimate achievements. Thus, it has three main components: setting appropriate goals, recording and verifying success in achieving a goal, and rewarding the achiever. The goals are typically set by an awarding organisation such as The Scout Association which set down the activities and skills required to gain badges of differing category and level. The traditional means of recording and verifying success is by a leader setting a series of tests and noting successful outcome. And the badges are visible, coveted indicators of achieving the award.

All these elements are now being automated and distributed among a community. For example, *iSpot* sets goals to achieve 'stars' in each category of wildlife based on the user's scientific expertise. The system automatically raises the profile of people who make many observations, who offer identifications that are confirmed by others, and who gain confirmed identifications from experts. A person who achieves a high level of expertise is not only rewarded visibly by up to five stars shown on their profile, but also carries more weight when confirming other people's observations.

The Mozilla Open Badge Infrastructure enables any organisation or community to define the skills and competences needed to achieve a badge. Each badge has a visible online image and a set of metadata that provides information about what the badge represents and the evidence used to support it (MozillaWiki, 2014). The Badge Backpack is a repository for each person to view a collection of badges, determine who should access them, and share them with others. A superficially similar system to record and recognise learning activities comes from the Tin Can API project. Its aim is to assist people in easy logging and storage of their learning activities wherever these occur – online or in the real world – in a standard format of "person did this" (e.g. "Sally experienced hang gliding", "Ian completed a FutureLearn course"). Additional fields provide information about the context and result of the activity.

While both Mozilla Open Badge and Tin Can API store and display elements of learning journeys, they differ in intent: the former is a record of achievement while the latter is a record of learning activity. This raises the intriguing prospect of combining both to create a composite record of a learning journey and its achievements (e.g. "Sally experienced hang gliding and obtained an open proficiency certificate", "Ian completed a FutureLearn course and gained an open statement of attainment"). A working group on 'Tin Badges/Open Can' is exploring just such a fusion of standards, to create shareable records of learning journeys. While these link activities and achievements, there still remains a need to record and share learning episodes and projects within a common framework (see Table 3.1).

A paper by Botički and colleagues (2014) describes a badge system for mobile learning. The *SamEx* application runs on devices with the Windows phone 7 and 8 mobile operating system in a Singapore primary school. The school students are encouraged to collect audio recordings, video clips and pictures on their mobile devices. They accumulate points by collecting these media, answering questions related to the location, providing comments on the work of other students and liking their work. These map onto a set of online badges that are awarded for accumulations of points in each activity.

Table 3.1 Examples of recording and enhancing elements of learning

Learning element	*Record*	*Enhance*
Achievement	Mozilla Open Badge	Mozilla backpack
Project	ePortfolios	LinkedIn
Episode	ePortfolios	
Activity	TinCanAPI	iSpot links

A study of 350 children using the application over one year identified three categories of student: 'badge hunters' who collected badges regardless of the quality of their contributions; 'dodgers' who were not interested in collecting badges at all; and 'sharers' who wanted to share with their peers. What was lacking from the application was a way to reward collaborative as well as individual learning.

Learners' dispositions and attitudes to learning are important factors influencing whether a successful learning journey will take place. In recent research on informal learning in professional contexts, a study of commercial airline pilots highlights the role of personal characteristics that have been found to motivate their informal learning, namely: initiative, self-efficacy, love of learning, interest in their profession, and professionalism (Corns, 2014). These are qualities that many teachers would also love to see and to develop in their students, for the learners' immediate benefit as well as in preparation for lifelong learning and work.

Conclusions

In this chapter we show how learning can be sustained through learning journeys, since these give an extended sense of continuity and progress. We have reflected upon learning journeys old and new. Mobile and ubiquitous technologies can implement a seamless continuity of learning within and across contexts, offering additional ways of combining learning in the classroom, at home and outdoors. It would appear that, so far, the goal of recording, sharing, assessing and celebrating learning journeys has been approached piecemeal with differing methods, standards and technologies. While some individuals appear to have a natural lifelong craving for learning, others need to be encouraged, supported and rewarded. To some extent, this mirrors the fragmented nature of education into a variety of formal, non-formal and informal settings each with their own requirements and methods of validation. But there is indication that it may be possible to create a sustainable record of learning in all its varied richness that can be captured on a variety of devices, curated by individual learners, groups, or organisations, and associated with transferrable badges to record achievement.

The notion of a learning journey continues to inspire educators and researchers and commonly manifests itself in mobile learning through the creation of location-specific tours and trails. It serves the important purpose of focusing attention on the value of mobile learning experiences in terms of real-world learning, across contexts, incorporating reflection on learning and having the potential to extend and sustain the learning according to personal or shared interests and needs. The notion of a learning journey also serves to establish a more transparent and mutually supportive relationship between formal and informal learning. Learning experiences may be actively supported across a range of diverse settings and there is growing recognition that informal elements can often be valued and rewarded.

References

Abbas Petersen, S., Procter-Legg, E., & Cacchione, A. (2014). LingoBee: Engaging mobile language learners through crowd-sourcing, *International Journal of Mobile and Blended Learning*, 6(2), 58–73.

Bartlett, C., Marshall, M., & Marshall, A. (2012). Two-eyed seeing and other lessons learned within a co-learning journey of bringing together indigenous and mainstream knowledge and ways of knowing. *Journal of Environmental Studies and Sciences*, 2(4), 331–340

Botički, I., Seow, P., Looi, C-K., & Baksa, J. (2014) How can badges be used in seamless mobile learning? *Proceedings of the Bristol Ideas in Mobile Learning 2014 Conference*. Bristol: University of the West of England. Retrieved on December 7, 2014 from http://cloudworks.ac.uk/cloud/view/8605

Bransford, J. D., Brown, A. L., & Cocking, R. R. (2000). How people learn: Brain, mind, experience, and school. Washington, DC: National Academy Press.

Cajete, G. (1994). *Look to the mountain: An ecology of indigenous education*. Asheville, NC: Kivaki Press.

Chan, T-W., Roschelle, J., Hsi, S., Kinshuk, Sharples, M., Brown, T. et al. (2006). One-to-one technology-enhanced learning: an opportunity for global research collaboration. *Research and Practice in Technology-Enhanced Learning, 1* (1), 3–29.

Clow, D., & Makriyannis, E. (2011, February). iSpot analysed: participatory learning and reputation. In *Proceedings of the 1st international conference on learning analytics and knowledge* (pp. 34–43). New York: ACM.

Corns, K. M. (2014). *Examining informal learning in commercial airline pilots' communities of practice*. Capella University dissertation. Retrieved on December 7, 2014 from http://gradworks.umi.com/36/15/3615070.html

DeFalco, A. (2010). An analysis of John Dewey's notion of occupations: Still pedagogically valuable? *Education and Culture*, *26*(1), 82–99.

Gaved, M., Jones, A., Kukulska-Hulme, A. & Scanlon, E. (2013). A citizen-centred approach to education in the smart city: Incidental language learning for supporting the inclusion of recent migrants. In: *New literacy and competences for Smart City learning*, Special issue of the *International Journal of Digital Literacy and Digital Competence*, 4(3), 50–64.

Gemmell, J., Bell, G., & Lueder, R. (2006). MyLifeBits: A personal database for everything. *Communications of the ACM*, *49*(1), 88–95.

Gordon, N. (2013). *Flexible pedagogies: Technology-enhanced learning*. York: The Higher Education Academy.

Hannon, P. D., Collins, L. A., & Smith, A. J. (2005). Exploring graduate entrepreneurship: A collaborative, co-learning based approach for students, entrepreneurs and educators. *Industry and Higher Education*, *19*(1), 11–23.

Hattie, J. (2009). *Visible learning: A synthesis of over 800 meta-analyses relating to achievement*. Abingdon: Routledge.

Hiemstra, R. (2001). Uses and benefits of journal writing. *New Directions for Adult and Continuing Education*, *2001*(90), 19–26.

Hong, J. (2013). Considering privacy issues in the context of Google Glass. *Communications of the ACM*, *56*(11), 10–11.

Hwang, G-J., Wu, P-H., & Ke, H-R. (2011). An interactive concept map approach to supporting mobile learning activities for natural science courses. *Computers & Education*, *57*(4), 2272–2280.

Jin, L., & Cortazzi, M. (2011). The changing landscapes of a journey: Educational metaphors in China. In: Janette Ryan (Ed) *Education reform in China: Changing concepts, contexts and practices*. Abingdon: Routledge.

Kukulska-Hulme, A., Gaved, M., Brasher, A., Jones, A., Scanlon, E., & Paletta, L. (2012). Designing for inclusion through incidental language learning. Conference paper at ICT for Language Learning. Florence, Italy, 15–16 November 2012.

Lave, J., & Wenger, E. (1991). *Situated rearning: Legitimate peripheral participation*. Cambridge: Cambridge University Press.

Martin, M. (2010). *Learning by wandering: An ancient Irish perspective for a digital World*. Oxford: Peter Lang.

Miller, J. (1996) *The holistic curriculum*. Toronto, Canada: OISE Press.

MozillaWiki (2014). Badges/Onboarding-Issuer. Retrieved on March 11, 2015 from https://wiki.mozilla.org/Badges/Onboarding-Issuer

Ogata, H., Houb, B., Li, M., Uosakic, N., Mouri, K., & Liu, S. (2014). Ubiquitous learning project using lifelogging technology in Japan. *Educational Technology & Society*, *17*(2), 85–100.

Palmer, Parker J. (2004) *A hidden wholeness: The journey toward an undivided life*. San Francisco, CA: Jossey-Bass.

Park, C. (2003). Engaging students in the learning process. *Journal of Geography in Higher Education*, *27*(2).

Reynolds, R., Speight, C., & Walker, K. (2008). Bridging formal and informal learning using mobile digital museum trails. 3rd WLE Mobile Learning Symposium, 27 March 2009, London. Retrieved on March 11, 2015 from http://www.lkl.ac.uk/people/kevin/wle_reynolds.pdf

Sharples, M., FitzGerald, E., Mulholland, P., & Jones, R. (2013). Weaving location and narrative for mobile guides. In: K. Drotner & K.C. Schrøder (Eds.), *Museum communication and social media: The connected museum*. New York & London: Routledge.

Sharples, M., Lonsdale, P., Meek, J., Rudman, P.D., & Vavoula, G.N. (2007). An evaluation of Myartspace: A mobile learning service for school museum trips. In: A. Norman & J. Pearce (Eds.) *Proceedings of 6th Annual Conference on Mobile Learning*, mLearn 2007. Melbourne: University of Melbourne, pp. 238–244.

Song, Y., Wong, L-H., & Looi, C-K. (2012). Fostering personalized learning in science inquiry supported by mobile technologies. *Educational Technology Research & Development*, *60*(4), 679–701.

Sprake, J. (2011). *Learning-through-touring: Mobilising learners and touring technologies to creatively explore the built environment*. Rotterdam: Sense Publishers.

Sutch, D. (2005) 'Bossing adults and finding spotty bras': Learners as producers within mobile learning contexts. Mlearn 2005 Conference, Cape Town, South Africa, 25–28 October 2005. Conference abstracts. Retrieved on March 11, 2015 from http://www.mlearn.org.za/CD/BOA_p.63.pdf

Tan, E.B.K., So, H-J., & Zhang, X. (2012). Teacher agency and student autonomy in inquiry-based mobile learning trail. Paper presented at the 20th International Conference on Computers in Education (ICCE 2012), Singapore, 26–30 November 2012.

Tsui, M.L.N., Tsui, E., & See-To, E.W.K. (2013). Adoption of a personal learning environment & network (PLE&N) to support peer-based lifelong learning. *Proceedings of The Asian Conference on Society, Education and Technology*, Osaka, Japan.

Tulving, E. (1985). Memory and consciousness. *Canadian Psychology/Psychologie Canadienne*, 26, 1–12.

Vavoula, G., & Sharples, M. (2002). KLeOS: A personal, mobile, knowledge and learning organisation system. *Proceedings of IEEE International Workshop on Wireless and Mobile Technologies in Education* (WMTE 2002), pp. 152–156.

Vavoula, G., & Sharples, M. (2009). Meeting the challenges in evaluating mobile learning: A 3-level evaluation framework. *International Journal of Mobile and Blended Learning*, *1*(2), 54–75.

Vavoula, G., & Sharples, M. (2009). Lifelong learning organisers: Requirements for tools for supporting episodic and semantic learning. *Educational Technology & Society*, *12*(3), 82–97.

Wheeler, M.A., Stuss, D.T., & Tulving, E. (1997). Toward a theory of episodic memory: The frontal lobes and autonoetic consciousness. *Psychological Bulletin*, *121*(3), 331–354.

Wong, L-H., & Looi, C-K. (2011). What seams do we remove in mobile-assisted seamless learning? A critical review of the literature. *Computers & Education*, *57*(4), December 2011, 2364–2381.

Worswick, L., Little, C., Ryan, K., & Carr, E. (2014) Interprofessional learning in primary care: An exploration of the service user experience leads to a new model for co-learning. *Nurse Education Today*. Retrieved on December 7, 2014 from http://dx.doi.org/10.1016/j.nedt.2014.05.007

4 Professional development for sustaining a mobile learning-enabled curriculum

Chee-Kit Looi, Daner Sun, Longkai Wu and Lung-Hsiang Wong

Introduction

Nowadays, curriculum innovations that integrate the use of ICT have been regarded as an essential strategy for educational reform. In Singapore, a series of policy statements were released to drive the integration of ICT into teaching and learning in every school (MOE, 2008). For example, the Third Masterplan (mp3) for ICT in Education that took place between 2008 and 2013 initiated the expansion and deepening of ICT integration in the curriculum. As one of the significant research projects responding to mp3, the seamless learning study aims to design and develop an innovative curriculum supported by mobile technology for primary school students. An innovation typically brings along challenges emanating from or faced by different parties or participants in the implementation process. The most significant factor lies with the teacher agents who act as the major practitioners (Fevre, 2014). Many reform initiatives for educational improvement did not accomplish the intended changes because they failed to change teachers (Fullan, 2007). Thus, a large body of research endeavours to develop high-quality Teacher Professional Development (TPD) opportunities for the teachers, as TPD is the key to building teacher capacity and implementing the innovation (Fishman et al., 2013).

In this seamless learning project, concurrent with the curriculum development, an ongoing TPD program has ensued. The intention was to support teachers in better understanding and implementing the curricular innovation in their teaching practices. If we are to facilitate the professional development of teachers, we must understand the process by which teachers grow professionally and the conditions that support and promote that growth (Clarke & Hollingsworth, 2002). In this chapter, we will detail the model of TPD in supporting the PD and teachers' curriculum implementation. We first review the literature about teacher changes and TPD, and the evidence-based developmental model for developing the theoretical framework for TPD. We then introduce the innovative curriculum, namely, Mobilized 5E-Science Curriculum (M5ESC). By articulating the major features of the curriculum, the pedagogical principles and ways of integrating it into the standard science curriculum will be highlighted. This will illuminate why a stage-by-stage and long-term TPD is necessary in the context of science learning. After presenting how the TPD was carried out over a period of five years in the school, we will investigate the effective features of the TPD through analysing data collected from the PD working sessions, teachers' interviews and classroom observations.

Literature review

Links between teacher changes and TPD

Researchers agreed that many educational innovations did not live up to their promises because the needs for teacher learning and teacher changes to adapt the innovation were not addressed adequately. There have been a flurry of discussions about teacher changes in knowledge and beliefs but less about changes in teaching practices (Bakkenes, Vermunt, & Wubbels, 2010). Studies have found that traditional TPD neglected the school contexts, teachers' needs, and the alignment between the intended pedagogy and teachers' real teaching practices, which led to the failure of teachers' growth, and further hampered the education reform efforts (Barone et al., 1996). While TPD may be part of the intervention for facilitating teacher changes, not all TPD is effective (Kwok, 2014).

A number of studies on TPD suggested that evidence is needed to establish the link between TPD and teacher practices that act as mediator between TPD and the instruction experienced by the students (Avalos, 2011; Brand & Moore, 2011; Dori & Herscovitz, 2005). Further, the research advocates that effective TPD for bringing about lasting change requires (a) new idea acceptance; (b) understanding of new practice integration; (c) sharing complex, tacit knowledge; (d) external support for change; and (e) sustaining community to facilitate change (Baker-Doyle & Yoon, 2011; Duke, 2004). In particular, reflective thinking skills and social interaction are highlighted as two components of the successful TPD. In the ICT-supported projects, how and why TPD works to support teachers' teaching decisions in the classrooms (Graham, Borup, & Smith, 2012), their compatibility with teachers' pedagogical beliefs and the technology being used, and the teacher change in perceptions and the pedagogical use of technology, have been regarded as the key determinants of curriculum success (Donnelly, McGarr, & O'Reilly, 2011). Thus, to evaluate the effectiveness of TPD, more efforts should be placed on investigating teachers' performance in teaching practices and their involvement in TPD, and more evidence should be sought to explore the mutual relationship between TPD and teachers' performance.

Fullan et al. (2005) pointed out that understanding the change process is a big driver in educational reform. Such understanding of the change process is about establishing the conditions for continuous improvement in order to persist and overcome inevitable barriers to reform. The evidence-based developmental model serves the purpose of gathering evidences from a staged-based curriculum innovation to establishing the connection between consecutive stages. The evidence captured is especially beneficial to practitioners for understanding the change process of the curriculum reform and for assisting them to implement the innovation (Klinger, Boardman, & McMaster, 2013). Meanwhile, evidence collected at each stage of the curriculum development informs the development of ongoing TPD. The evidence is an important mediator between TPD and teacher practices.

M5ESC innovation

Pedagogical principles of M5ESC

The development and implementation of M5ESC was our preliminary attempt to systematically and comprehensively explore the integration of mobile learning into a science curriculum for a long-term and stage-by-stage intervention. It was developed iteratively and progressively via design-based research in a Singapore primary school since 2008 (Looi et al., 2009). The curriculum is mapped to national science curriculum standards, and covers all of the standard materials at P3 and P4 required in a primary school. M5ESC is aligned with the learning objectives in the Primary Science Syllabus (MOE, 2007, pp. 2–3) and Singapore Ministry of Education (MOE)'s advocacy on the development of twenty-first century competencies in science education. It aims to promote students' conceptual understanding of science and develop critical learning skills (e.g. collaborative learning skills, self-directed learning skills, reflective thinking skills) (MOE, 2010; Sha et al., 2012). Several key elements were incorporated into the curriculum in order to achieve these goals (Thijs & van den Akker, 2009). Table 4.1 depicts the key elements and features of the current M5ESC.

Science teaching in the K–12 classroom is often posited in science education reform documents to be centred on inquiry (NRC, 1996). The 5E (Engagement-Exploration-Explanation-Elaboration-Evaluation) instructional model has been pervasively employed in the instruction of science in Singapore schools (Bybee, 2006). In the M5ESC environment (Figure 4.1), guided by the structured inquiry framework, students can take the lead in the inquiry phases for developing inquiry skills and self-directed learning skills (Crawford, 2007). In the classroom, constructivist learning theory supports inquiry by placing the focus of learning on student ideas, questions, and understanding, and not teacher delivery of content (Fosnot, 1996). Teachers are encouraged to apply constructivist teaching approaches to ask questions, conduct mobile and non-mobile activities, interact with students and scaffold the students' learning. Equipped with mobile devices, individual students may carry out the learning activities at their own pace and pursue their preferred learning paths. The M5ESC learning activities

Table 4.1 Key elements of M5ESC

Key Elements	*M5ESC*

Pedagogical Principles	• Design lesson plan based on 5E instructional model
	• Instruction using constructivist pedagogical approaches:
	  ◦ Ask exploratory and open-ended questions
	  ◦ Conduct student-centred activities
	  ◦ Provide scaffolding to students' inquiries

Content of Subject	• Singapore Primary Science Curriculum at P3 and P4
Learning Objectives	• Develop knowledge, understanding and application of key scientific concepts, skills and processes, ethics and attitudes as mentioned in the Primary Science Syllabus. • Emphasise the development of twenty-first century competencies: i.e. self-directed learning skills, collaborative learning skills and reflective thinking skills.

Resources	• Learning resources
	  ◦ Science textbook
	  ◦ Activity worksheet
	  ◦ School-based worksheet
	  ◦ Mobile device with MyDesk system
	  ◦ Multimedia and Internet resources
	• Teaching resources
	  ◦ Teacher guide book
	  ◦ Science textbook
	  ◦ Mobile device with MyDesk authoring tools
	  ◦ Multimedia resources

Learning Activities	• Mobile learning activities
	  ◦ Classroom hand-on activities
	  ◦ Sharing and discussion activities
	  ◦ Experiments
	  ◦ Field trip and home activities
	• Non-mobile learning activities
	  ◦ Classroom hand-on activities
	  ◦ Sharing and discussion activities
	  ◦ Experiments
	  ◦ Field trip and home activities

Location/Time	• Classroom: in class time • Out of Classroom: out of class time

Teacher's Role	• Support flexible roles in the classroom
	  ◦ Lesson designer and practitioner
	  ◦ Facilitator of students' activities
	  ◦ Guide for the field trip activities
	  ◦ Evaluator of students' performance

Student's Role	• Support learners as active learners
	  ◦ Self-directed learners
	  ◦ Collaborators in group work
	  ◦ Evaluators of learning artefacts

Table 4.1 Key elements of M5ESC (Continued)

Key Elements	*M5ESC*

Technology's Role	• Improve students' conceptual understanding and skills via Mydesk
	  ◦ Self-reflection of prior knowledge and knowledge changes
	  ◦ Conceptual mapping of scientific concepts
	  ◦ Relating knowledge with daily life experience
	• Monitor students' progress and learning process

Assessment	• Formative assessment
	  ◦ Activity worksheet
	  ◦ MyDesk learning artefacts
	  ◦ Activity performance
	• Summative assessments

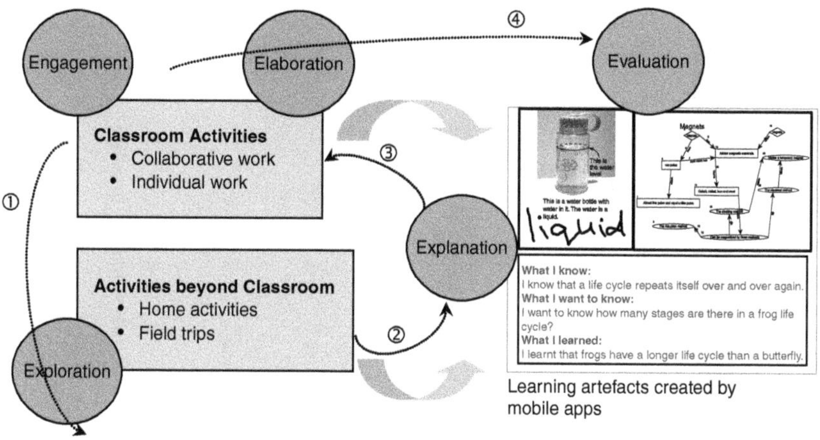

Figure 4.1 The M5ESC environment

can be conducted both in and out of the classroom according to the principles of seamless learning (Wong & Looi, 2011).

MyDesk integration in the M5ESC

To enable and enhance the students' seamless learning experience and outcomes, we developed the MyDesk system as an integral part of the M5ESC environment. The MyDesk system is the multifunctional tool that supports teachers' lesson design and students' inquiry activities. In M5ESC, the MyDesk system that runs on a Microsoft Windows Mobile operating system is flexibly integrated with the 5E inquiry phases. Using the MyDesk teacher portal (Figure 4.2a), the teachers create learning tasks/activities for the 5E inquiry-based lessons that require the use of multiple media and

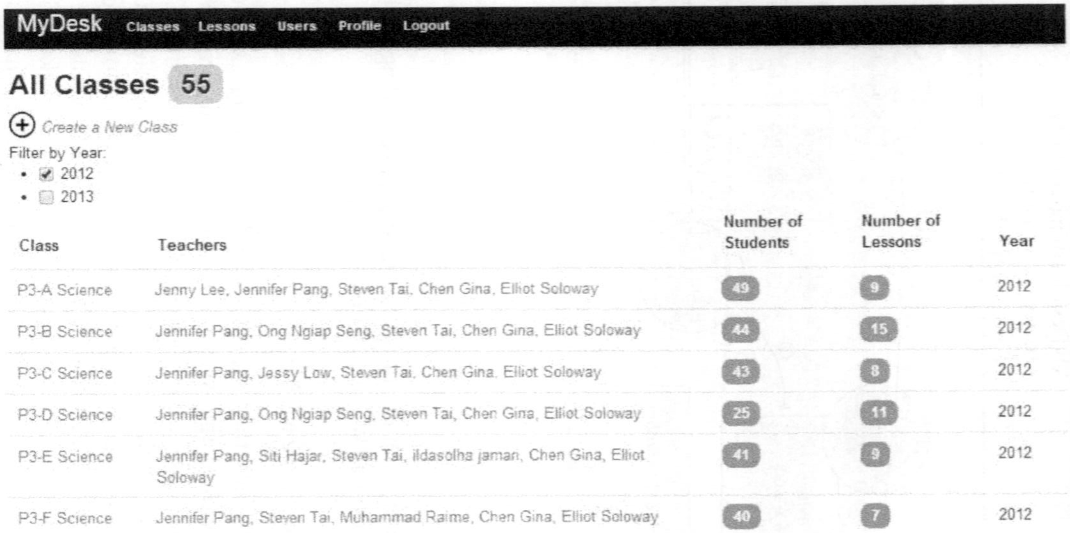

(a)

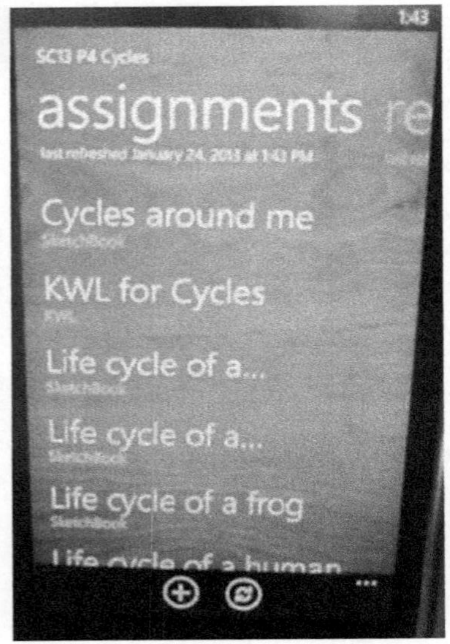

(b)

Figure 4.2 (a) The MyDesk teacher portal (b) The MyDesk students' module

Table 4.2 The learning tools of MyDesk learning system

Tools	*Functions*	*Mobile activities in Fungi*
KWL	• A self-reflection tool supporting students' reflecting upon on learning process and conceptual changes through responding questions (i.e. what do I already **K**now? what do I **W**ant to know? What have I **L**earned?) to allow students to learn in a self-regulated way.	• Engagement: students respond to "what do I already know" about fungi in KWL. • Exploration: students respond to "what do I want to know" about fungi in KWL. • Evaluation: students respond to "What I have learnt" about fungi in KWL.
Sketchbook	• An animation/drawing and picture annotating tool to assist students' establishing connections between knowledge learned in the classroom and knowledge applied outside the classroom.	• Engagement: students record the changes of moist bread and toasted bread using Skecthbook.
MapIt	• A concept map tool that allows students to develop conceptual understanding through creating, sharing, and exploring concept maps.	• Elaboration: students draw concepts maps of the characteristics of fungi using MapIt.
Blurb	• A question setup tool which facilitates the teacher to set up specific questions to ask students to give short opinions or feedback on their inquiry activities or their understanding of knowledge.	• Exploration: students respond to the questions: how do the fungi grow? in Blurb.
Recorder	• A voice recorder tool for students to record the process of the experiment, fieldtrip and the observation of teacher demonstration, and students' reflection and conclusion are also recorded as a data for teachers' to review their progress and improvement in inquiry.	• Exploration: students record their questions when observing the moist and toasted bread using Recorder.
Notepad	• A data recording tool for students to record the results or process of experiments, fieldtrip, and observation of teacher demonstration.	• Engagement: students write their observations of the moist and toasted bread using Notepad.

applications (e.g., text, graphical, spreadsheet, animations, and the like), and then review and comment on students' work generated in the activities (Looi et al., 2009). Students can assess the learning activities and complete their tasks using learning tools in the students' module of MyDesk (Figure 4.2b). The learning tools in MyDesk include KWL (carrying out self-reflection through responding to "what I know" and "what I want to know" before, and to "what I learnt" after learning the topic), NotePad (taking notes), Recorder (recording voice), Sketchbook (drawing models), MapIT (constructing concept maps), and Blurb (posing questions). These tools are intended to facilitate students in developing sophisticated and systematic understanding of scientific concepts, enhancing skills in modelling, reasoning and reflective thinking, and especially fostering self-directed learning in and out of the classroom (Brooks & Brooks, 1993; Greca & Moreira, 2000).

Table 4.2 depicts the learning tools and their functions, and some exemplar mobile learning activities in the lesson unit of Fungi in P3 science.

Complementing these researcher-developed tools are the affordances offered by both the device and applications available on the web, such as mobile blogs, online discussion forums, video/photo cameras, and search engines. With this suite of tools, teachers may evaluate student artefacts by giving a grade, and posting feedback and comments.

Lesson exemplar

Table 4.3 presents the lesson design of "Exploring Materials" in Primary 3 science. We delineate activities both in and out of classroom with a specific focus on how the technology was integrated at different phases of the 5E model.

The design of M5ESC is aligned with the vision that learning needs to be more individualised, learner-centred, situated, collaborative, ubiquitous, and continuous. The incorporation of different tools facilitates the blending of a variety of learning activities within one learning environment (Naismith et al., 2005). In M5ESC, more time is allocated to the instruction of mobile learning activities compared with the non-mobile activities. Teachers and students interact more frequently in the sharing and discussing of work generated by mobile tools. Formative assessment is treated as an integral part of the learning process and an important source for students and teachers to make reflections on learning and teaching (Nicol & Macfarlane-Dick, 2006). Indeed, students' performance at mobile learning activities, which present their involvement in the learning process, is an important indicator of their learning gains throughout the learning process of M5ESC. In summary, the M5ESC seeks to provide a new avenue for teachers to conduct learner-centred mobile learning activities. It helps students to reinforce their autonomy of doing learning activities and to refine their understanding in various learning settings, thereby stimulating their motivation towards science learning (Cullen, Harris, & Hill, 2012).

Table 4.3 Design of activities for "Exploring Materials"

Lesson sequence	*Activities*	*Technology integration*
Lesson 1: Engagement	Classroom activity: Teacher demonstrates the properties of materials (e.g. hardness)	Camera, Internet search
	Home activity: Students draw concept maps of material classification	MapIT
Lesson 2: Exploration	Classroom activity: • Teacher reviews and identifies inappropriate conceptions of materials classification in KWL; • Students collaboratively design and conduct experiments to explore the properties of materials.	Camera, Notepad, KWL
	Home activity: • Students take pictures of materials in their daily life and describe their properties and usefulness; • Write their thoughts and reflection in KWL.	Camera, video camera, KWL, Sketchbook
Lesson 3: Explanation	Classroom activity: Teacher identifies misconceptions in KWL and Sketchbook; and guides students to improve their understanding.	KWL, Sketchbook
	Home activity: Students design and conduct experiments independently to demonstrate properties of materials in their daily life.	Camera, video recorder, Notepad, KWL
Lesson 4: Elaboration	Classroom activity: • Students present their work done at home; • Students peer-critique and share their ideas; • Teacher explores and stimulates students' conceptual understanding by further explanation and summary.	KWL, Sketchbook
Lesson 5: Evaluation	Classroom activity: Students apply their understanding to answer worksheet questions.	–

Evolution of the TPD model for M5ESC

A proper, well-planned TPD program is needed to ensure that innovative curricula be introduced in a sustainable manner (Bell & Gilbert, 1996). As mentioned above, M5ESC is a curricular innovation that integrates inquiry-based instructional principle and combines the use of different mobile learning tools for facilitating teachers to create a constructivist learning environment for the students to learn science in a self-directed way. A TPD model needs to be evolved concurrently for leveraging up teacher competencies in the context of such comprehensive curriculum innovation.

The literature on TPD mentioned that an effective TPD should not only have a structural and evolved developmental process, but also have its emphasis or strategies for supporting the developmental phases. An optimal TPD should afford:

1 The reciprocal relationship between TPD and accountability in that the teachers must give consent to what they are being asked to do and the strategies for doing it (Elmore, 2002). TPD is not a short term working session, but an interactive and recursive continuum with complex reactions among multiple factors (Clarke & Hollingsworth, 2002).
2 A learning community for promoting teacher discussion, peer assessment and self-reflection, particularly those focusing on curriculum and instruction (Scribner et al., 1999). The focus is not just on individual teacher's professional learning but also on professional learning within a community context.
3 Situated learning catering to teacher needs. As the effective TPD models indicate, the successive stages with different emphases and forms of activities facilitate the development of teacher competencies.

Thus, the TPD for M5ESC is intended to mirror these key features of optimal TPD. In a later section, we describe the developmental process of the ongoing TPD over a period of five years (from 2009–2013). Typically, as an institutional response to external challenges, a rational planning approach to change initiative, which comprises elements such as need analysis, research and development, strategy formation, resource support, implementation and dissemination, and evaluation (Lueddeke, 1999), will be effectuated. Such a systematic approach is certainly helpful to identify actual concerns and to engage teachers and stakeholders in the change practices (Yuen, 2004).

To facilitate teachers' adoption and adaptation of the M5ESC, a TPD developmental process followed by an action research model was put in place and has since evolved. Action research is a form of investigation designed for use by teachers to attempt to solve problems and improve professional practices in their own classrooms. It involves systematic observations and data collection which can be then used by the practitioner-researcher in reflection, decision-making and the development of more effective classroom strategies (Parsons & Brown, 2002). A long-term and stage by stage TPD modality is preferred (Bradshaw, 2002) over extensive numbers of one-time TPD workshops or programs. Hence, the duration of TPD for M5ESC is consistent with the development and implementation of the M5ESC from 2009 to the present. The development of TPD is a long-standing category of activity and programs rooted in the school context and teachers needs with different stages (Carroll et al., 2005). Figure 4.3 depicts the development stages of TPD for M5ESC.

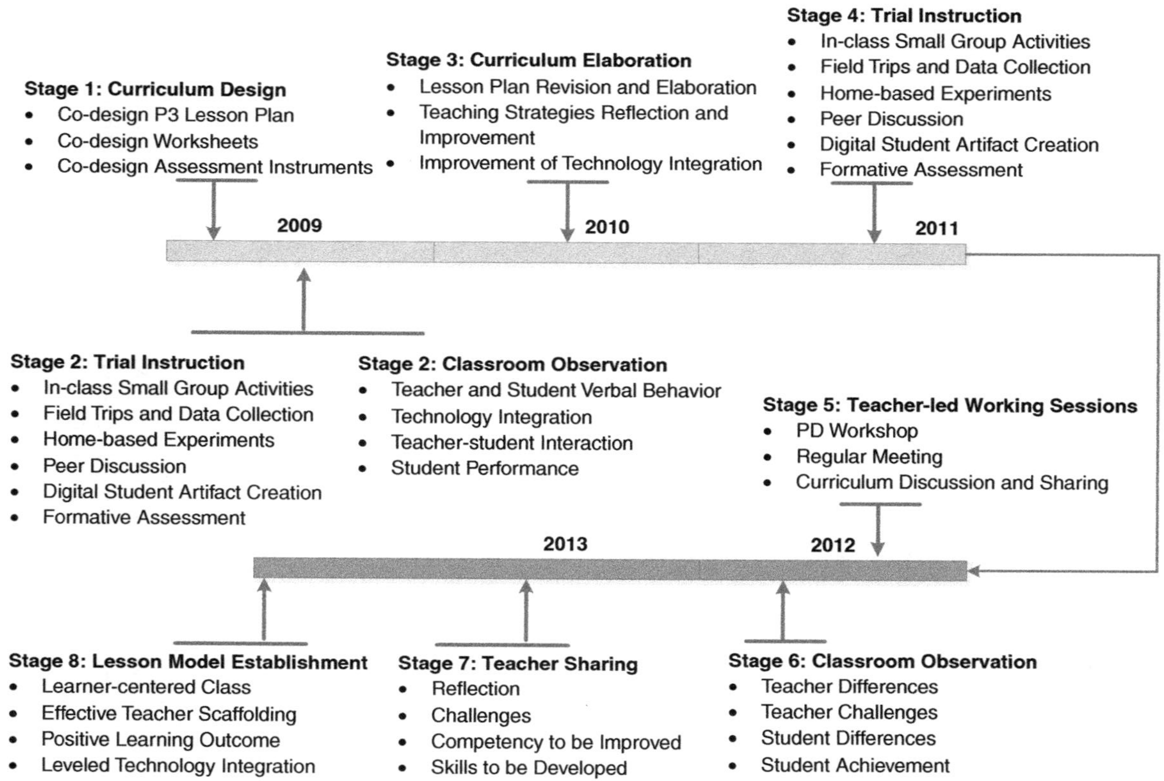

Figure 4.3 The historical trajectory of teacher professional development for M5ESC

Our TPD model advocates the evolution of TPD connected with teachers' instructional practices, the collaboration between researchers and teachers on curriculum development and elaboration, and teachers' active involvement in decision-making (Day, 1999; Orlando, 2013). The TPD comprises two continuous phases with various parties' engagement (see Figure 4.3): Phase I: Curriculum elaboration and trial implementation phase (stages 1–4) and Phase II: Curriculum scaling phase (stages 5–8). Phase I attempts to establish the exemplar lessons for M5ESC, attain consensus on the pedagogical principles and relevant teaching strategies, and nurture a leading teacher for the consequent Phase II. Phase II emphasises the development of teachers' competencies in teaching, thus shifting their pedagogical beliefs about the mobile technology-facilitated curriculum (Ballone & Czerniak, 2001). Thus TPD can be traversed in an iterative way, offering possibilities to teachers to return to the previous stages to revisit and further reflect on the experiences and knowledge therein, for the subsequent stages.

At stage 1, a leading teacher (Jodie,[1] head of the science department with six years of teaching experience) was selected to co-design the curriculum materials with researchers in 2009. Upon completion of lesson planning for six topics at P3 level, the lessons were enacted by Jodie in her mixed ability class.[2] Two researchers observed the lessons with special attention on teacher and students' verbal behaviours, the patterns of teacher-student interactions, the organisation of student activities, the way in which technology was integrated, and student actions and performances (An & Reigeluth, 2012; Inan et al., 2010). Meanwhile, timely feedback and comments regarding design and enactment were provided by researchers after each lesson to facilitate Jodie's lesson reflection. Students' artefacts posted to MyDesk were also assessed as another indicator of teacher performance on the lesson enactment. The findings informed the curriculum elaboration at stage 3, where different parties including educators, teachers, and researchers sat together to revise the lesson design, reflect on teaching practices, share analytical findings, and discuss the gaps between the designed and the enacted lessons in the school year of 2010. To seek further alignment between the enactment and the purposes of the curriculum, another round of implementation and observation of the refined lessons was carried out at stage 4 in 2011.

With positive outcomes of the pilot run of the M5ESC for three school years, the school decided to scale up the M5ESC from one class to all eight classes of Primary (Grade) 3 in 2012. Apart from Jodie, five other teachers joined the teaching force of M5ESC.[3] To support teacher adaptation of M5ESC pedagogy, a teacher learning community in the form of teacher-led working sessions comprising of the newly-joined teachers, the subject head, ICT head, curriculum planner, and researchers was formed to meet on a weekly basis at stage 5. We found this factor to be influential in establishing ownership, which would positively influence the teachers' approach to curriculum development and their level of engagement. As a leading teacher, Jodie provided valuable experiences and knowledge on the enactment of M5ESC, the principles of organising learning activities, the support and scaffolding provided to the students, and the patterns of integrating technology in the classroom based on constructivist pedagogical approaches. In the working sessions, they reviewed and revised the enacted lesson plans. They also discussed and sought consensus on the proposed teaching strategies for the forthcoming science lessons. In particular, they discussed how the curriculum could be customised for students of different abilities and yet at the same time retain the design intent of the curriculum.

Additionally, workshops targeting at developing teachers' Pedagogical Content Knowledge of 5E, inquiry-based instruction, and in and out-of-classroom mobile technology use were conducted to assist new teachers for better alignment of their lesson enactment and the lesson design. The researchers made numerous visits to classrooms during implementations to support teachers and observe the implementation and its fidelity to what was intended in the designed lesson plans. Moreover, some novice teachers' classes were selectively observed by experienced teachers together with researchers (Cuckle & Clarke, 2002). At stage 6, a fine-grained data analysis was conducted to compare teacher and class differences, and to identify the difficulties and challenges encountered by the teachers in their respective classes.

Teacher sharing sessions were regularly held for discussing the problems identified and seeking possible improvement on the learning process and scaffold designs, as informed by teachers' reflections and researchers' observation and feedback. Specifically, the research team emphasised critical pedagogical reflections about teaching as an integral part of the TPD (Penso, Shoman, & Shiloah, 2001). In the TPD sessions, the teachers negotiated possible and better solutions to the tensions between the traditional assessment and the formative assessment of M5ESC.

The TPD model in our study is not only built on the features of good TPD models in the literature, but also active and frequent interactions between researchers' practices, teachers' practices and students' responses. The execution of one stage is followed by reflections upon previous practices or evidence collected during long-term TPD process. Through participating in the TPD for M5ESC, teachers are expected to draw towards the intended pedagogy with good fidelity. Ultimately, the TPD evolves together with the teacher's growth in curriculum development and instruction, and students' progression in achievement.

The teacher learning community

Roles of participants

In education, most systemic transformation efforts involve stakeholders that are critical to achieving the desired changes (Watson, Watson, & Reigeluth, 2008). The M5ESC curriculum innovation went through different

Table 4.4 Roles of different participants at different stages

Phase	*Teachers*	*Students*	*Researchers*	*Principal*	*Parents*
Design	Designer	Experimenter	Designer	Supporter	Supporter
Implementation	Practitioner Analyser	Practitioner	Observer, Analyser	Reviewer	Supporter
Assessment	Evaluator Reflector	Evaluator Reflector	Evaluator Analyser	Reviewer	Supporter

development stages wherein different participants were involved in the co-design, co-implementation and co-assessment of the curriculum and innovation from their own perspectives. In M5ESC, the shifts in the roles of participants contributed to the achievement of curriculum development and TPD at various stages. Five distinctive participants were identified as the major contributors to bridge the gap between the desired curriculum and the enactment curriculum. Table 4.4 describes the roles of teachers, students, researchers, principal and parents at the different stages of M5ESC development.

Teacher changes have been regarded as the key determinants of curriculum success (Donnelly, McGarr, & O'Reilly, 2011). Before teachers can confer a sense of empowerment to students, they must first feel themselves to be strong and engaged professionally and personally (Dierking & Fox, 2012). Farrell and Weitman (2007) propose teacher empowerment as three-pronged: teachers experiencing more access to decision making, achieving greater knowledge, and seeing improved teacher status. The TPD for M5ESC provides such environment for the teachers to empower their instruction. In addition to serving in an instructional role, the experienced teachers acted as mentor, supervisor, counsellor, and community leader of the teacher learning community. At the same time, they served as the designer of the curriculum, as well as analyser of their own lessons and others' lessons. Finally, their self-assessment and peer assessment would provide more information on how to elaborate the curriculum and teaching practices. In M5ESC, researchers acted as the mediators among these participants in that they co-designed the lessons, observed the teaching practices, and proposed suggestions for teachers to improve their teaching in and out of TPD activities. More importantly, the researchers sought and obtained evidence on students' performances to correlate with different strategies or enactments from teachers. Students were also encouraged to voice their comments in their experiences of M5ESC. Moreover, the principal provided full support for the M5ESC implementation and scale up at more grade levels; milestone meetings were conducted to share his perspectives and his views on the classroom changes. Parents were encouraged to support the students' learning in their homes and in informal contexts; their feedback on students' performance in the informal learning was valuable as feedback to improve the curriculum. Thus, while sharing the common perspective that transforming the traditional curriculum into a constructivist-based mobile curriculum, the evolvement of M5ESC was closely entwined with the mutual interactions and co-efforts among different stakeholders.

Characterising the TPD working sessions

Key components of working sessions

As mentioned above, all P3 teachers participated in the TPD working sessions with the researchers from 2012. The working session consists of three components, namely, research sharing and discussion; teacher reflection and sharing; and lesson improvement and elaboration (Table 4.5).

This kind of teacher learning community particularly deepened and refined the shared vision related to the teaching of technology-supported activities. The effective use of technology and the teacher learning community are mutually supportive in that intensity and depth of engagement in one create conditions for increases in the other (Dexter, Seashore, & Anderson, 2002).

Nature of interactions in TPD working sessions

To analyse the involvement of teachers in the TPD working sessions, researchers did the video and audio recording of each session. In this study, 9 sessions which focused on the discussion of curriculum elaboration and implementation were selected (4 sessions from 2012 TPD, and 5 sessions from 2013 TPD). The discourse from all 9 sessions was first coded into the categories delineating the major essence of the discussion, namely, "principles and pedagogy", "feedback and reflection", "lesson design", "challenges and problems", and "lesson plan". The researchers also identified the foci and frequency for each category in order to explore the factors for enhancing or inhibiting the teacher enactment of the lessons. The specific contents of each category are as follows:

1 Principles and pedagogy: this refers to the introduction, discussion, sharing of the principles and pedagogy concerning the design and implementation of M5ESC. Researchers elaborated on the principles and pedagogies of M5ESC, from which teachers shared and discussed their understanding of these principles and pedagogies. Teachers also discussed how to apply the shared principles and pedagogies. Teachers would have practical sessions where they applied the intended principles and pedagogies. For instance, before the weekly TPD working session, the researchers had assigned some readings on inquiry questioning to all the teachers involved. When the researchers and teachers sat together, some volunteer teachers role-played with the researchers to practise the questioning techniques using the content from the school syllabus. Through

Table 4.5 The key components of the teacher learning community

Components	*Research sharing and discussion*	*Teacher reflection and sharing*	*Lesson improvement and elaboration*
Purpose	• To promote teacher understanding of the pedagogical principles of M5ESC and relevant teaching strategies.	• To encourage teachers' critical analysis of their instructional practices to identify and understand factors influencing their lesson enactment (Brand & Moore, 2011).	• To improve teacher competency on the lesson design and implementation.
Content	• Researchers introduce learning theories and pedagogical approaches (e.g. 5E inquiry-based instruction, collaborative learning, self-directed learning, reflective thinking, and mobile learning and seamless learning) that back the design and implementation of M5ESC as required by the teachers. • Teachers share their understanding and discussed the application of these pedagogical principles in teaching practices.	• Teachers share and reflect upon their actual enactments of the designed lesson. • Teachers reflect upon and discussed students' responses with a focus on identifying students' misconceptions and the teaching strategies to address these misconceptions. • Teachers share and discuss teaching strategies to address the learning needs and difficulties of students with different ability levels.	• Teachers re-design instructional materials and learning activities. • Teachers elaborate the teaching strategies catering to different ability students.

the processes of role-play, the volunteer teachers as well as the observing teachers gained more authentic experience in using that questioning technique. This helped teachers to build their confidence and competencies in applying pedagogical techniques in real instructional situations.

2 Feedback and reflection: Teachers provided feedback on their lesson enactment (i.e. the use of teaching and learning resources, the application of teaching strategies, students' responses, and parents' responses) and reflected upon the enactment. Researchers also shared their observations of and provided suggestions for teacher enactment. This sharing can also inform the enhancement of the curriculum. Take the activity of a zoo field trip for example. After the implementation, teachers reflected in the TPD working session that the zoo trip was in general beneficial to students as they gained more real experience which helped them to recall and relate the intended scientific concepts. Yet as some students (especially the low ability students) were observed to have some difficulty learning in-situ, the teachers suggested adding some pre-activities to get students better prepared for field learning (e.g. showing students the pictures of the zoo animals taken by previous cohorts of students, teaching students how to do observation and documentation of their observations).

3 Lesson Design: Discussions focused on the new design or re-design of learning activities, mobile learning activities, methods of content delivery and assessment, differentiated instruction of activities, and guiding questions. Based on lesson enactment feedback and reflection, teachers and researchers re-designed and elaborated the formal and informal learning activities, instructional materials, and assessments to address the learning needs and gaps of students of different ability levels. For example, when teachers found that low-ability students had some problems identifying the characteristics for doing classification of products in supermarket, they proposed some scaffolding (e.g. guiding questions) to be incorporated.

4 Challenges and Problems: Discussions focused on time issues, students' readiness of mobile learning activities and parents' feedback. Teachers provided feedback on the technical (e.g. poor network connections) and logistical issues (e.g. lack of instructional time due to school events, need for technical training for both teachers and students), and discussed the possible work-around or solutions with the researchers.

5 Lesson Plan: Discussions focused on the content and the design of the key teaching materials; the school-based lesson plan[4] and the relevant students' learning materials (i.e. topical worksheet, and activity worksheet).

The data analysis suggested that teachers paid more attention to sharing their feedback and reflection on the teaching strategies (n=54)[5] and lesson enactment (n=37) compared with other discussions, with the aim of elaborating the instruction of M5ESC (Figure 4.4). Meanwhile, they would like

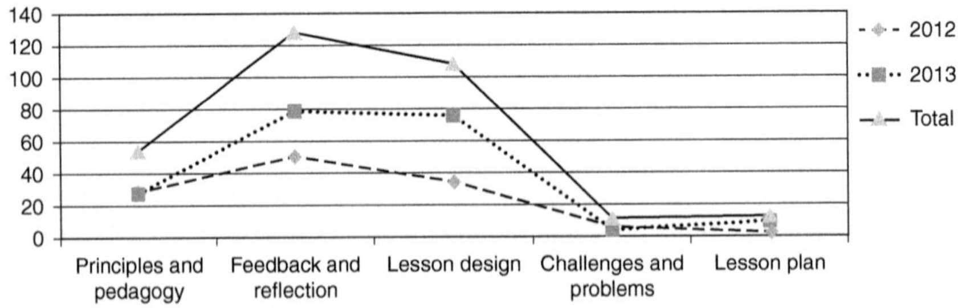

Figure 4.4. Various foci of discussions in the TPD sessions

to revise or rework the lesson design (i.e. introducing teacher demonstration with a guiding question as the alterative activity of watching video), especially for the design of mobile learning activities (n=40) (i.e. designing MyDesk learning activities). More specifically, they spent some time on discussing the task design, the technology integration, and the questioning design for the mobile learning activities in the informal settings (e.g. zoo, science garden, and supermarket). Moreover, in 2013, teachers were still keen to seek a deeper understanding of the principles and pedagogy in M5ESC. Most of them discussed further principles of constructivist questioning skills, reflection and sharing, ICT skills and inquiry skills. They also discussed the challenges and problems in the lesson design and lesson enactment. They finally commented and revised the original lesson plans for future enactment in the next school year.

In the following sections, we provide a discussion of the factors that contributed to teacher's good enactment of the M5ESC from their TPD working sessions. For this purpose, we select three TPD working sessions that have focused on the design of mobile learning activities, the design of outside classroom learning activities and the strategies on motivating students' involvement in learning. The discussion reflected teachers' understanding of M5ESC and their efforts on the elaboration of the M5ESC.

Feedback and reflection: the design and implementation of KWL learning activities

Most teachers encountered challenges in conducting the learning activities that took place outside the classroom (Solvberg & Rismark, 2012). In the TPD working sessions, when discussing the design of mobile learning activities, such as KWL activity, the teachers pointed out that the KWL should not been interpreted as the homework, although KWL activities were assigned as outside classroom tasks. From their reflections on students' KWL work, they mentioned that the KWL did not require correct or normative answers, and that it provided students with opportunity of reflecting on their learning process and conceptual understanding. Furthermore, the teachers agreed that the KWL served as a learning platform between the teacher and their students to communicate about their experiences on learning science, to share their ideas and to elaborate their knowledge. In the classroom, they shared the KWL postings empowering teachers to monitor and guide the students' outside learning activities. They saw that the assessment and sharing activities of KWL would provide more space for students to express their ideas and for the teacher to obtain valuable information on students' misconceptions and current understanding levels. The students became aware that their inputs and efforts were being acknowledged, and that they were contributing to the class discussions. This would further attract more students to be more involved in the reflection work, and to lead to deeper reflection (e.g. critical reflection) in their learning.

Lesson design: how to design out-of-classroom tasks?

In TPD working sessions, the teachers devoted great efforts to discussing the design of out-of-classroom tasks, for example, in the topic of Fungi, a Sketchbook task referring to the observation of fungi growth on moist and toasted bread. The task requires students to take pictures of changes in bread, as well as to describe and explain the phenomena. During the discussion, the teachers agreed with the revision and elaboration on the instruction of the task: "compare your observations between both breads, reason and explain your observation". They thought it would be valuable for students' to compare and explain their observation and describe their findings with image postings, which enabled teachers to capture more about students' thinking and knowledge.

Feedback and reflection: students' responses to the M5ESC

The teachers reflected that they gradually understood the value of exploratory questions in which one question could lead to another. The follow-up questions were generated and posed based on students' responses and ideas. Some teachers found that the students gained better motivation in answering questions and sharing their ideas. They reflected that before M5ESC, the students were given few opportunities to do experiential learning, students had few chances to do science; even as they took pictures and videos, they rarely shared them in the classroom and rarely engaged in peer discussion on the learning artefacts. In M5ESC, the teachers felt that students were engaged in both classroom activities and outside activities. The students enjoyed the learning experiences of capturing data (e.g. images, video, notes), and sharing their data in the classroom. Students became enthusiastic in the peer and classroom discussions. They benefited from the learning experiences through which they learnt knowledge in the classroom, applied this in the daily life and then elaborated on when back in the classroom. In particular, the M5ESC created more opportunities for lower ability students to share and improve their knowledge; the teachers could listen to them and guide them on further elaborating their knowledge. Through student work on the MyDesk platform, the teachers could have a good sense of their understanding levels, and could highlight the misconceptions arising from their work outside of the classroom.

Feedback and reflection: strategies on motivating students' engagement in learning

In the TPD sessions, the teachers proposed that more questions should be posed for students to ponder, and scaffolding in the form of hints and scripts should be provided. They suggested a more open learning environment should be established for students to respond freely without any concern about making mistakes. They pointed out that learning contexts should be created to motivate students' interests through connecting their knowledge learned in the classroom with knowledge application outside of classroom. The teachers felt more discussions and sharing should be conducted for students to learn from each other, and benefit more from their sharing, debating and negotiation process. They also suggested involving more students in sharing and discussion by providing more scaffolding of knowledge, skills and task purposes.

Through analysing the discussion in TPD sessions, we identify the factors that influenced the lesson enactment of teachers and the aspects in which they can further improve. This kind of TPD working session created a community-centred teaching learning environment and ensured that there was time for teachers to work together and to provide feedback to each other. More importantly, they improved their understanding of some principles, elaborated their teaching strategies and noticed students' positive changes impacted by M5ESC. Through deep reflection, the teachers received opportunities to process, renegotiate and share their prior understanding of relevant principles and teaching strategies, and to further recognise the appropriateness of the teaching strategies and methods.

Teacher voices

In this section, we provide a narration of some teachers' voices based on their experiences in the TPD sessions and in the enactment of M5ESC. We share the views of three participating teachers, namely, Tom, Alice, and Jemmy, whose views are representative of the views from all the participating teachers.

At the beginning of the M5ESC implementation, as Tom expressed low confidence in enacting the curriculum, he relied mostly on the textbook and teaching resources for preparing the lessons, and did not have a deep understanding of the principles of the M5ESC. After joining the TPD and doing the lessons, he developed more confidence in implementing the curriculum. He noted the changes in students that resulted from the autonomy given to them in doing their hands-on activities and their participation in peer discussion and sharing of their learning experiences. He found that students become more engaged in the discussion and sharing work:

> Now I have more confidence to conduct the activities from M5ESC. When I saw my students actively discussing their work and sharing their ideas with their partner, I understood the value of student-centred activities. I attempt to conduct more activities based on the co-designed lesson plans to develop students' skills in collaboration and communication. And I'm amazed that there are kids that conduct experiments or do activities at home even though they were the weaker lot and their parents were also quite supportive. Most students could post their work done out of classroom, and they shared their process with their classmates and peer assessment of each other's work; they enjoyed the learning process and seemed to be more engaged in the mobile learning activities compared to the paperwork activities.

Previously, Alice felt that the lack of her ability of doing technology integration was the obstacle for implementing mobile learning activities in the class. But later, she was able to apply technology in more activities with different cognition levels into the class (Starkey, 2011). During these activities, she valued students' participation and contribution to their own inquiry activities. For example, in the lesson of "Exploring Materials", the students developed deeper understanding of the properties and value of the materials and objects in their daily lives after they engaged in a series of activities using Sketchbook in MyDesk: looking for the products → taking the pictures → describing the constituent materials → pointing out the values and properties of the materials.

> But what I like about the phone is that the alternative platform allows children who may be reluctant to do the written work part… they at least have an alternative tool to get them to draw, to record, to crate and to generate their own work. The use of smartphone opens up my ideas on the technology use in more learning activities with different learning objectives. For example, I can teach students' skills on data collection using camera and audio recorders in the smartphone; and I can ask students to practice their reflective thinking skills using the KWL app, and assist students to develop system thinking skills through using MapIT app. These did not happen in my previous classroom.

Jemmy shared that his students now spent more time on learning from books, and as they pursued understanding through interacting with teacher and their classmates, they became more open to peers' experiences for better understanding. These changes emerged after he implemented M5ESC for around one year. The same changes were found in their ways of seeking for answers that they would rather do peer discussion with their partners and not directly approach the teacher. Jemmy felt that he now paid more attention to how to scaffold students' peer discussion while knowing more about students' prior knowledge and providing appropriate scripts or prompts for them to find the solutions.

> I guess I am using more inquiry-based teaching (strategies) in class. I am more conscientious with using the inquiry-based teaching in class. I probe more – usually when my children do not give the answer, I will just tell them the answer, like "hey this is it". But now is like I probe a little more, I ask more questions and I get them to think more. Interestingly, students were not eager to get the answers from me but discussed with their partners until they reached the same answers.

Conclusion

In this chapter, we have described the development and implementation of an evidence-based curriculum and the concomitant evolvement of its long-term and continuous TPD program. The TPD is a viable response to the importance of teacher role acting in the curriculum innovation. Our TPD is in resonance with the principles of effective TPD as reported in the literature, namely, that TPD should: (a) be intensive, ongoing, and connected to practice; (b) focus on student learning and address the teaching of specific content; (c) be aligned with school improvement priorities and goals; and (d) build on strong working relationships among teachers (Marrongelle, Sztajn, & Smith, 2013). TPD should take into account teachers' needs as learners (Brand & Moore, 2011), and incorporate an understanding of how teachers learn to transform their knowledge into practice for the benefit of their students' growth (Avalos, 2011).

In our TPD, the teacher practitioners played different roles, and their PD work was rooted in the school context in that they practised the innovation and learned from each other in a progressive manner. The TPD working sessions enabled teachers' self-reflection and sharing of their thoughts and experience for forming and sustaining an active learning community for both novice and expert teachers involved in the seamless learning project. The tracing of teachers' growth and their changes is rooted in long-term observations and analysis, and is not just based on episodes in a limited time frame. With these efforts, the innovation and its TPD have been shown to be effective for both teachers and students in the context of this primary school where the curricular innovation of M5ESC is enacted.

In summary, we refined the major features of our TPD for M5ESC as shown in Figure 4.5. A teaching and learning community including one

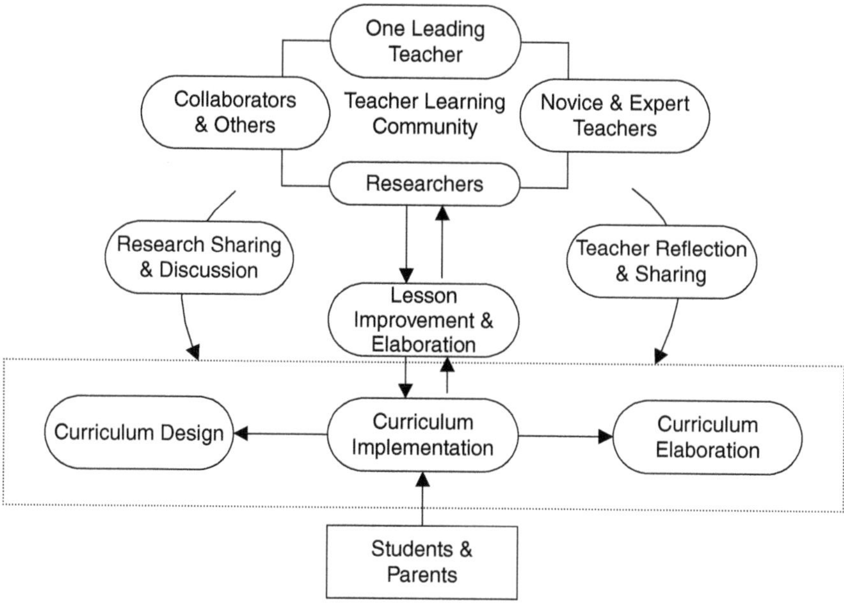

Figure 4.5 Major features of TPD for M5ESC

leading teacher, novice and expert teachers, researchers and other participants has evolved and grown since the beginning of the curriculum development and the establishment of TPD. The discussion on the contributing factors of TPD on curriculum design and implementation from research sharing and discussion, teacher reflection and sharing, and lesson improvement and elaboration reflected the major features of the effective TPD model. The model changes teachers into active learners, shaping their professional growth through reflective participation in TPD and in practice (Clarke & Hollingsworth, 2002). We recognised the importance of the teacher-led conversations and the interaction with the researchers in and out of TPD working sessions. We also paid attention to the feedback from the students and their parents, which impacted the lesson elaboration further. The TPD for M5ESC is a dynamic system designed to serve the desired goals of curriculum innovation and yet evolving through interactions among the different parties involved. As to further work, we will explore ways in which our TPD can be applied to a wider teacher community, which is concerned with teacher collaboration and growth across several schools. We envision such an evidence-based TPD model will be leveraged for the capacity building of teachers from different backgrounds and accounts, as well as bringing varied enactments and adaptations of the curricular innovation in different school contexts.

Acknowledgements

The paper is a part of work from the project "Bridging Formal and Informal Learning Spaces for Self-directed and Collaborative Inquiry Learning in Science", funded by Singapore National Research Foundation (NRF2011 – EDU002-EL005). We would like to thank our other team members Xie Wenting, Peter Seow, and Gean Chia, and teachers and their students from our pilot school for working on this project.

Notes

1 'Jodie' is a pseudonym to protect the identity and privacy of the participant – so are the rest of the names in this paper.
2 To support sustained learning activities, 34 students from the experimental class were each assigned a smartphone with 24x7 access in order to mediate a variety of learning activities such as in-class small-group activities, field trips, data collection and geo-tagging in the neighbourhood, home-based experiments involving parents, online information search and peer discussions, and digital student artefact creation, among others.
3 There were six teachers for these eight classes, and they included the teacher in our initial study. Four teachers started the implementation at the beginning of the academic year, while the other two teachers joined the implementation a few months later.
4 The lesson plan is designed based on the MOE syllabus and the pedagogical principles of M5ESC. The lesson plan has been iteratively revised and implemented along with the curriculum design and development since 2009.
5 n represents the frequency of the discussion in the specified category.

References

An, Y.-J., & Reigeluth, C. (2012). Creating technology-enhanced, learner-centered classrooms: K-12 teachers' beliefs, perceptions, barriers, and support needs. *Journal of Digital Learning in Teacher Education*, *28*(2), 54–62.

Avalos, B. (2011). Teacher professional development in Teaching and Teacher Education over ten years. *Teaching and Teacher Education*, *27*(1), 10–20.

Baker-Doyle, K. J., & Yoon, S. A. (2011). In search of practitioner based social capital: A social network analysis tool for understanding and facilitating teacher collaboration in a U.S.-based STEM professional development program. *Professional Development in Education*, *33*(1), 75–93.

Bakkenes, I., Vermunt, J. D., & Wubbels, T. (2010). Teacher learning in the context of educational innovation: Learning activities and learning outcomes of experienced teachers. *Learning and Instruction*, *20*(6), 533–548.

Ballone, L.M., & Czerniak, C.M. (2001). Teachers' beliefs about accommodating students' learning styles in science classes. *Electronic Journal of Science Education*, *6*(2), 1–43.

Barone, T., Berliner, D. C., Blanchard, J., Casanova, U., & McGowan, T. (1996). The future of American teacher education. In J. Sikula, T. J. Buttery, & E. Guyton (Eds.), *Handbook of research on teacher education* (2nd edition; pp. 1108–1149). New York: Macmillan.

Bell, B., & Gilbert, J. (1996). *Teacher development: A model from science education.* London: Falmer Press.

Bradshaw, L. K. (2002). Technology for teaching and learning: strategies for staff development and follow-up support. *Journal of Technology and Teacher Education*, *10*(1), 131–150.

Brand, B. R., & Moore, S.J. (2011). Enhancing teachers' application of inquiry-based strategies using a constructivist sociocultural professional development model. *International Journal of Science Education*, *33*(7), 889–913.

Brooks, J. G., & Brooks, M. G. (1993). *In search of understanding: The case for constructivist classrooms.* Alexandria, VA: Association for Supervision and Curriculum Development.

Bybee, R. W. (2006). *The BSCS 5E instructional model: Origins and effectiveness.* A report prepared for the Office of Science Education, National Institutes of Health. Retrieved on December 7, 2014 from http://sharepoint.snoqualmie.k12.wa.us/mshs/ramseyerd/Science%20Inquiry%201%2020112012/What%20is%20Inquiry%20Sciecne%20(long%20version).pdf

Carroll, J. M., Rosson, M. B., Dunlap, D., & Isenhour, P. (2005). Frameworks for sharing teaching practices. *Educational Technology & Society*, *8*(3), 162–175.

Clarke, D., & Hollingsworth, H. (2002). Elaborating a model of teacher professional growth. *Teaching and Teacher Education*, *18*(8), 947–967.

Clough, G., Jones, A. C., McAndrew, P., & Scanlon, E. (2008). Informal learning with PDAs and smartphones. *Journal of Computer Assisted Learning*, *24*(5), 359–371.

CPDD of MOE, (2007). *Science syllabus primary 2008* (pp. 2–3). Retrieved on December 7, 2014 from http://www.moe.gov.sg/education/syllabuses/sciences/files/science-primary-2008.pdf

Crawford, B. A. (2007). Learning to teach science as inquiry in the rough and tumble of practice. *Journal of Research in Science Teaching*, *44*(4), 613–642.

Cuckle, P., & Clarke, S. (2002). Mentoring student-teachers in schools: views, practices and access to ICT. *Journal of Computer Assisted Learning*, *18*(3), 330–340.

Cullen, R., Harris, M., & Hill, R. R. (2012). *The learner-centered curriculum: Design and implementation.* Retrieved on December 7, 2014 from http://www.moe.gov.sg/education/syllabuses/sciences/files/science-primary-2008.pdf

Day, C. (1999). *Developing teachers: The challenges of lifelong learning* (p. 11). London: Routledge.

Dexter, S., Seashore, K. R., & Anderson, R. E. (2002). Contributions of professional community to exemplary use of ICT. *Journal of Computer Assisted Learning*, *18*(4), 489–497.

Dierking, R. C., & Fox, R. F. (2012). Changing the way I teach: Building teacher knowledge, confidence, and autonomy. *Journal of Teacher Education*, *64*(2), 129–144.

Donnelly, D., McGarr, O., & O'Reilly, J. (2011). A framework for teachers' integration of ICT into their classroom practice. *Computers & Education*, *57*(2), 1469–1483.

Dori, Y. J., & Herscovitz, O. (2005). Case-based long-term professional development of science teachers. *International Journal of Science Education*, *27*(12), 1413–1446.

Duke, D. (2004). *The challenges of educational change.* New York, NY: Pearson.

Elmore, R. F. (2002). *Bridging the gap between standards and achievement: The imperative for professional development in education.* Washington, DC: Albert Shanker Institute.

Farrell, J. B., & Weitman, C. J. (2007). Action research fosters empowerment and learning communities. *Delta Kappa Gamma Bulletin*, *73*(3), 36–45.

Fevre, D. M. L. (2014). Barriers to implementing pedagogical change: The role of teachers' perceptions of risk. *Teaching and Teacher Education*, *38*, 56–64.

Fishman, B., Konstantopoulos, S., Kubitskey, B. W., Vath, R., Park, G., Johnson, H., & Edelson, D. C. (2013). Comparing the impact of online and face-to-face professional development in the context of curriculum implementation. *Journal of Teacher Education*, *64*(5), 426–438.

Fosnot, C. T. (1996). Constructivism: A psychological theory of learning. In C. T. Fosnot (Ed.), *Constructivism: Theory, perspectives, and practice* (pp. 8–33). New York: Teachers College Press.

Fullan, M. (2007). *The new meaning of educational change*. New York, NY: Teachers College Press.

Fullan, M., Cuttress, C., & Kilcher, A. (2005). 8 forces for leaders of change. *Journal of Staff Development*, *26*(4), 54–64.

Graham, C. R., Borup, J., & Smith, N.B. (2012). Using TPACK as a framework to understand teacher candidates' technology integration decisions. *Journal of Computer Assisted Learning*, *28*(6), 530–546.

Greca, I. M., & Moreira, M.A. (2000). Mental models, conceptual models, and modelling. *International Journal of Science Education*, *22*(1), 1–11.

Inan, F. A., Lowther, D. L., Ross, S. M., & Strahl, D. (2010). Pattern of classroom activities during students' use of computers: Relations between instructional strategies and computer applications. *Teaching and Teacher Education*, *26*(3), 540–546.

Kember, D., Leung, D., Jones, A., & Loke, A.Y. (2000). Development of a questionnaire to measure the level of reflective thinking. *Assessment and Evaluation in Higher Education*, *25*(4), 380–395.

Klingner, J.K., Boardman, A.G., & McMaster, K.L. (2013). What does it take to scale up and sustain evidence-based practices? *Exceptional Children*, *79*(2), 195–211.

Kwok, P.-W. (2014). The role of context in teachers' concerns about the implementation of an innovative curriculum. *Teaching and Teacher Education*, *38*, 44–55.

Looi, C.-K., Wong, L.-H., So, H.-J., Seow, P., Toh, Y., Chen, W., ... Soloway, C. N. (2009). Anatomy of a mobilized lesson: Learning my way. *Computers & Education*, *53*(4), 1120–1132.

Lueddeke, G.R. (1999). Toward a constructivist framework for guiding change and innovation in higher education. *The Journal of Higher Education*, *70*(3), 235–260.

Marrongelle, K., Sztajn, P., & Smith, M. (2013). Scaling up professional development in an era of common state standards. *Journal of Teacher Education*, *64*(3), 202–211.

Ministry of Education. (2008). MOE launches third masterplan for ICT in education. Retrieved on December 7, 2014 from http://www.moe.gov.sg/media/press/2008/08/moe-launches-third-masterplan.php.

Ministry of Education. (2010). MOE to enhance learning of 21st century competencies and strengthen art, music and physical education. Retrieved on March 23, 2015 from http://www.moe.gov.sg/media/press/2010/03/moe-to-enhance-learning-of-21s.php

Motiwalla, L. F. (2007). Mobile learning: A framework and evaluation. *Computers & Education*, *49*(3), 581–596.

Naismith, L., Lonsdale, P., Vavoula, G., & Sharples, M. (2005). Literature review in mobile technologies and learning. A Report for NESTA Futurelab. Available from NESTA FutureLab.

National Research Council. (1996). *The national science education standards.* Washington, DC: National Academy Press.

Nicol, D. J., & Macfarlane-Dick, D. (2006). Formative assessment and self-regulated learning: a model and seven principles of good feedback practice. *Studies in Higher Education*, *31*(2), 199–218.

Orlando, J. (2013). ICT-mediated practice and constructivist practices: is this still the best plan for teachers' uses of ICT? *Technology, Pedagogy and Education*, *22*(2), 231–246.

Parsons, R.D., & Brown, K.S. (2002) *Teacher as reflective practitioner and action researcher*. Belmont, CA: Wadsworth/Thomson Learning.

Penso, S., Shoman, E., & Shiloah, N. (2001). First steps in novice teachers' activity. *Teacher Development*, *5*(3), 323–338.

Rau, P.-L. P., Gao, Q., & Wu, L.-M. (2008). Using mobile communication technology in high school education: Motivation, pressure, and learning performance. *Computers & Education*, *50*(1), 1–22.

Rodríguez, P., Nussbaum, M., & Dombrovskaia, L. (2012). ICT for education: a conceptual framework for the sustainable adoption of technology-enhanced learning environments in schools. *Technology, Pedagogy and Education*, *21*(3), 291–315.

Ruchter, M., Klar, B., & Geiger, W. (2010). Comparing the effects of mobile computers and traditional approaches in environmental education. *Computers & Education*, *54*(4), 1054–1067.

Scribner, J. P., Cockrell, K. S., Cockrell, D. H., & Valentine, J. W. (1999). Creating professional communities in schools through organizational learning: An evaluation of a school improvement process. *Educational Administration Quarterly*, *35*(1), 130–160.

Sha, L., Looi, C.-K., Chen, W., Seow, P., & Wong, L.-H. (2012). Recognizing and measuring self-regulated learning in a mobile learning environment. *Computers in Human Behavior*, *28*(2), 718–728.

Shih, J.-L., Chuang, C.-W., & Hwang, G.-J. (2010). An inquiry-based mobile learning approach to enhancing social science learning effectiveness. *Educational Technology & Society*, *13*(4), 50–62.

Solvberg, A., and Rismark, M. (2012). Learning spaces in mobile learning environments. *Active Learning in Higher Education*, *13*(1), 23–33.

Starkey, L. (2011). Evaluating learning in the 21st century: a digital age learning matrix. *Technology, Pedagogy and Education*, *20*(1), 19–39.

Thijs, A., & van den Akker, J. (Eds.). (2009). *Curriculum in development*. Enschede, Netherlands: SLO – Netherlands Institute for Curriculum Development. Retrieved on December 7, 2014 from http://www.slo.nl/downloads/2009/curriculum-in-development.pdf/

Tobin, K., Kahle, J.B., & Fraser, B.J. (1990). *Windows into science classrooms: Problems associated with higher level cognitive learning in science*. London: Falmer Press.

Watson, S.L., Watson, W.R., & Reigeluth, C.M. (2008). Systems design for change in education and training. In J. M. Spector, M. D. Merrill, J.J.G. van Merrienboer & M. P. Driscoll (Eds.), *Handbook of Research on Educational Communications and Technology* (3rd ed.). Mahwah, NJ: Lawrence Erlbaum Associates.

Wildy, H., & Wallace, J. (1995). Understanding teaching or teaching for understanding: Alternative frameworks for science classrooms. *Journal of Research in Science Teaching*, *32*(2), 143–156.

Wong, L.-H., & Looi, C.-K. (2011). What seams do we remove in mobile-assisted seamless learning? A critical review of the literature. *Computers & Education*, *57*(4), 2364–2381.

Yuen, H.K. (2004). Leading curriculum innovation in primary schools. *New Horizons in Education*, *49*, 99–107.

5 Sustaining mobile learning at a personal level

Mobile digital literacy

Wan Ng

Introduction

Mark Weiser first articulated the concept of Ubicomp, or ubiquitous computing, in relation to three waves of computing (Weiser, 1991). The first wave saw the use of one mainframe computer by many people; the second wave saw a one-to-one computer to human ratio, where individuals each owned a desktop or laptop. We are now in the third wave of computing where one individual is served by many computing devices, for example a mobile phone, digital camera, global positioning system (GPS), a games console, and for some people a computer tablet, laptop and desktop as well.

In education, institutions are focusing on the one-to-one (1:1) concept, where the hope is that every student has access to an individual computing device in the classroom. For the last decade, many schools have focused on laptop or notebook programs where each student purchased, loaned or was provided with one of these devices by the schools. For example, in Australia, between the years 2009 and 2013, schools were funded by the *Digital Education Revolution* initiative (Australian Information and Communications Technology in Education Committee [AICTEC], 2009) to ensure that all year 9–12 students had access to a laptop. As the initiative came to an end, the bring-your-own-device (BYOD) alternative began attracting the attention of education departments and institutions. Globally, BYOD is gaining in popularity in schools and higher education institutions. BYOD is a technology model where students bring a personally-owned smart mobile device (e.g. laptop, netbook, smartphone or computer tablet) from home to school for the purpose of learning. Such devices are able to connect to the Internet and contain apps for learning. BYOD is attractive for several reasons. It is a way of learning that uses technologies the students already possess and are familiar with because they make use of them in their everyday lives. This way of working has become easier because in the last few years the evolution of smartphone and tablet technology has made them more powerful in terms of their processing power and capabilities. In particular, there is a plethora of educational apps available across all discipline areas to assist students with their learning. With the prices of mobile handheld devices falling, the ownership of smart mobile devices by adolescents and young adults is increasing at a rapid pace. The Ofcom (2013) report indicated that smartphone ownership by 12–15 year olds in the United Kingdom is at 62%, and that there has been a tripling of the number of tablet computers in the homes of 5–15 year olds – a sharp increase from their 2012 survey. A United States Pew Internet Research (2012) study found a similar trend of 66% smartphone ownership by young adults aged 18–29 years. The research anticipated that almost all adolescents and young adults will own a smart device in the next few years, with the ownership starting at a much younger age. In Australia, a 2013 research project (Ng & Nicholas, manuscript in preparation) on 1,129 years 7 to 10 students (12–15 year olds) found that 97% of the students owned a mobile phone, with two-thirds (66%) of this group of adolescents owning smartphones and more than two-thirds (71%) owning or having access to a tablet. Other studies that show an increase in tablet ownership in families include a recent report on a study called *Zero to eight: Children's media use in America, 2013* (Rideout, 2013). It found that 75% of the children eight years and under had access to a smart device such as smartphone and/or tablet. The study reported a fivefold increase in tablet ownership in families with children in this age group, to 40%, up from 8% in 2011. The research also found that the percentage of children under two years of age using mobile devices for media has increased from 10% in 2011 to 38% in 2013 and that children from lower-income families have increased their access to mobile devices, from 22% in 2011 to 65% in 2013. The smartphone ownership pattern among young people in Asia is also similar, where ownership by 15–29 year olds has increased from 66% to 98% in Taipei, Singapore, Hong Kong and Seoul (Odaka, 2013).

While learners around the world do not have equal access to computers, especially those in developing countries, many have access to mobile phones. The 2013 International Telecommunication Union reported that mobile phone subscriptions worldwide numbered more than 6.8 billion at the end of 012 (ITU, 2013), with the developing countries now being more mobile than the developed world (Kelly and Minges, 2012). The access to mobile phone technologies by people in developing countries is enabling researchers to investigate ways of delivering education via mobile phone technologies to the people in these countries. For example, University of Cambridge researchers in collaboration with China Mobile were examining the potential of mobile phones to deliver *mHealth*, i.e. healthcare in China and worldwide (University of Cambridge, 2010). Other researchers such as del Carmen Valderrama Bahamondez and Schmidt (2011) were researching ways of overcoming the scarcity of computers in schools in developing countries. They asserted that the accessibility of mobile phones could change this situation and bridge the digital divide by providing the potential to build feasible educational applications that would enhance the learning experiences of students in these countries.

The capabilities of mobile phones, even those with enhanced features, differ from smartphones and tablets, which are able to access the Internet. Despite the limited capabilities of mobile phones, careful planning of learning activities in using them, such as through the use of SMS and mobile phone camera, has been shown to enhance learning (Ekanayake & Wishart, 2011; Keogh, 2011; Thornton & Houser, 2005). Other social/pedagogical uses of mobile phones based on text messages have been reported in projects such as *SMS4Learn*[1] (an initiative using SMS in Afghanistan to support training) and *The Jokko Initiative –Senegal*,[2] which involves collaborative work between UNICEF and Tostan in West Africa, aimed at encouraging group decision-making in Senegalese villages through the use of mobile phone text messages.

Having access to a personal mobile smart device that is able to access the Internet enables a student to access resources and interact with people online in an 'always on' capacity, where the student is able to expand his/her knowledge base enormously through a combination of formal and informal learning. Mobile devices enable knowledge acquisition in context and on a needs basis, making learning more meaningful and relevant to the student. What the educator omits in the classroom teaching on a topic of interest could be followed up with research on the Internet in the student's own time and space. Indeed mobile learning could be redefined according to the ubiquity of mobile devices. Trucano (2014) argues that the meaning of 'mobility' depends on an individual's perspective. He states that from an observer's perspective, the mobile device moves as it is carried by the user from place to place. From the user's perspective, the mobile device is a constant, kept in a place (pocket, bag, table) but locked in the physical relationship to the user providing virtual access to other worlds that are changing constantly around it. Trucano (2014) questions whether it is the device and the user that move through the world or the world that moves around them, mediated through the small screen on the mobile device. For many young people today, the latter is more evident since everywhere one looks, they are often interacting more with their devices than with the real world around them. In the physical classroom, interaction is guided by the educator, where academic interaction is usually with textbooks or e-content on computer devices and social interaction is with the educator and peers through discussions and group work. In a hypothetical virtual classroom, the interactions are literally boundless – interacting with the huge amount of information in different formats and modes on the Web and communicating with the large number of online communities associated with a wide variety of interests.

While the mobile education field is expanding, crucial issues underpinning practices and their sustainability remain to be addressed. A central issue is a pedagogical framework that connects formal and informal learning enabled by mobile technology. Education institutions are also required to address the sustainable use of mobile devices for learning in relation to factors such as policy, infrastructure and technical support, the role of educators and the type of professional development required to prepare them for teaching in mobile technology-enhanced environments. There are also issues associated with students' learning with mobile devices to be considered. For example, Chan et al. (2006, p. 27) state that despite the positives that one-to-one and mobile computing offer for learning, there is a

> need to remain cognizant of the potential negative issues associated with one-to-one TEL (technology enabled learning)... namely, the penetration of pervasive computing into all of one's life-spheres with potential for creating an unbalanced lifestyle; data security, integrity and privacy issues; persistent digital divide; high environmental and ecological costs of low-cost pervasive computing; and learning supported by one-to-one TEL for unethical and socially destructive purposes.

In the literature, there are some studies that explicitly mention teaching the students how to use mobile devices and/or a particular mobile app for a given project, but in general, there appears to be an underlying assumption that learners have the necessary competencies to learn with mobile technologies. Little about the literacy associated with interacting and learning with mobile technologies has been written or incorporated into the frameworks of research studies. Raising awareness of the issues mentioned by Chan et al. (2006), and teaching students the skills and knowledge to handle them are part of developing *mobile digital literacy* in students. Well-developed mobile digital literacy empowers students to learn more effectively with mobile technology as well as function better in everyday life activities that often rely on the capabilities of mobile devices to solve their everyday problems. The aim of this chapter is to present mobile digital literacy and the interacting components within it as a means of clarifying how to sustain mobile learning at the personal level. Sustainability in this context is defined as the continuing use of mobile technologies for safe and effective learning across the individual's lifespan.

Sustainability of mobile learning supported by mobile digital literacy is significant for two key reasons. First, the current generation of students have more ownership and easier access to mobile technologies (computer tablets, games consoles and smartphones) than their previous generations of peers. For many of the students, access to mobile technologies begins at a very early age, often pre-school age. Since access to mobile technologies has increased dramatically in families, by the time the children start schooling, many would have explored technology and developed technological and social skills through playing games, chatting with grandparents and relatives on *Skype* or *Facetime*, watching videos and interacting with educational software. Early childhood educator Jenn Rogers cited in Saylor (2011) called this "street" knowledge that educators should capitalise on for learning. The implication is that students come into the classroom with some degree of mobile digital literacy and a set of skills that are largely associated with social networking, Internet searching for general information, entertainment and gaming. These skills are developed largely through "tinkering" and exploring individually or with friends and family members (Ito et al., 2008). The second reason is that despite the increased interactions with technologies, research has shown that the young people are far less fluent than is generally assumed with standard digital tools such as web browsers and word processing programs and have difficulty in using the academic Web, such as searching for information, identifying resources, understanding the multimodal aspects of information, categorising, summarising and evaluating Web-based information (Tally, 2007). In addition, other researchers (e.g. Helsper & Eynon, 2010; Kennedy et al., 2008; Margaryan, Littlejohn, & Vojt, 2011; Sánchez et al., 2011) have indicated that students who have grown up with technology (the Generation Ys, millenials and digital natives) do not necessarily know how to make use of technology effectively for learning. In Ng (2012a) I argued that the purposeful use of technology by young people in informal settings is not characterised by exploring educational technologies but rather consists of social networking where there is value and purpose in keeping in touch with friends and family, making more friends and finding out about the world through their social networks. I therefore argued that it is the role of educators to teach young learners about learning with technology and to raise awareness about the types and flexibility of available applications that could be used for learning. Educators are therefore challenged to help students reverse unfavourable habits that they may have developed in their informal development of digital literacy, for example their underdeveloped web-based search-and-assess skills and the use of digital content ethically for academic purposes. As most of the informal development of digital literacy is with mobile devices, this chapter builds on the mlearning literacy described by Ng (2013) and proposes mobile digital literacy as a subset of digital literacy where the constraints of using mobile devices for learning and everyday activities are considered. It argues for the development of skills and competencies associated with mobile digital literacy to sustain the individual to learn and solve everyday problems in a safe and effective manner.

Mobile digital literacy

In my paper, 'Conceptualising mlearning literacy' (Ng, 2013), I have argued specifically for learning with mobile devices that is curriculum focused. The mlearning literacy concept is somewhat narrow for today's young people and students who have been immersed in a mobile-mediated world from a very young age (Rideout, 2013). In a young person's life, learning that is mobile-mediated is not confined to the classroom or curriculum content. Students make use of mobile technology for a variety of activities on a daily basis. They learn new things each time they explore a new mobile application (app), be that a bank app, an offline map (e.g. *MapsWithMe*) or a game app. They learn new things when they engage in conversations on blogs or social network sites. They expand their knowledge and conduct research spontaneously, in an on-demand manner with their mobile devices, for example, reading up more about a person or event to distinguish between facts and fiction after watching a movie. Being able to use mobile technology effectively for a wide variety of activities on a daily basis for learning and/or other everyday activities requires skills and knowledge about the mobile device, how to use it to solve problems, how to interpret the data shown on the screen and to make decisions based on the data. These are mobile digital skills and competencies that will be referred to as *mobile digital literacy* in this chapter. Prior to discussing the mobile digital literacy framework, it is appropriate to define mobile learning and discuss the uses of mobile devices in education and everyday living within the context of the framework.

Mobile learning and mobile devices

Mobile learning refers to the use of handheld (hence mobile) devices for the delivery of education. It encompasses the processes of pedagogical design of instruction and its implementation by educators to the learning by students that leads to achieving desired goals and learning outcomes. In order to teach and learn with mobile devices, educators and students need to acquire technical control of the device and understand its limitations at both the hardware and software levels. For the educator, knowledge of how to search for suitable apps and assess their suitability is also required. In a society where having some control over the authoring of apps to adapt them to one's use is encouraged, being able to code or re-code apps would be an advantage. School-based technology curriculum documents are beginning to call for the teaching of digital technology for deeper understanding of the underlying principles such as "use computational thinking and information systems to define, design and implement digital solutions" as stated in the Australian Curriculum (ACARA, website). Similarly the *National curriculum in England: Computing programmes of study* (Department of Education, 2013) national curriculum sets as a goal that all students "can understand and apply the fundamental principles and concepts of computer science, including abstraction, logic, algorithms and data representation and can analyse problems in computational terms, and have repeated practical experience of writing computer programs in order to solve such problems".

The mobile devices referred to in this chapter are handheld tools that are in some ways constrained in their processing capacity and include the mobile phone, smartphone and tablets but exclude laptops and notebooks. Hence, this chapter makes the case that while there are similarities in the digital literacies associated with mobile devices and laptop/desktop usage, the ease in portability and connectedness coupled with the constraints in size and processing power of mobile devices make mobile digital literacy a specific and differentiated construct. Nevertheless, Houghton (2013, online) contended that "the mobile computer increasingly fuses tools for the comprehensive multimedia presentation of ideas with tools for their creation and communication" and asserted that mobile devices are thought-making machines of the twenty-first century that are portable everywhere. Similarly, Low (2007, p. 79) described today's mobile devices as "converging many tools into one, leading to some items like mobile phones being coined the 'Swiss Army Knife' of this century" with their strength lying in their convenience and pervasiveness. The communicative and social-enabling aspects of mobile devices are also strengths that promote learning and ease problem-solving tasks. In order to discuss mobile digital literacy for learning and problem solving, it is necessary to discuss these and other affordances that mobile devices offer for learning, and the theories underpinning the learning.

Mobile devices as social learning tools

A significant part of Web 2.0 is its social web with online tools that allow people to learn by:

- networking socially e.g. through blogs, *Facebook*, *Facetime, Google+*, *Twitter* and other social platforms – providing that their mobile devices are Internet connected, students can use the facilities listed above to discuss projects and homework with their peers and sometimes with their teachers;
- reading, writing and editing in collaborative knowledge construction activities e.g. through wikis, *GoogleDocs* and *cMaps*;
- sharing ideas and information, e.g. through blogs, slideshare, podcasts (via *iTunes* or *uTunes*), videos on *YouTube* or by contributing comments/opinions to online newspapers and/or other discussion forums.

The increased capacity to network with more learners and experts via Web 2.0 together with the 'always on' capability of mobile devices mean that learners are able to access more resources and interact more frequently synchronously and/or asynchronously with others to enhance their learning. Nevertheless, there is no single way of understanding how learning occurs in such environments. Various learning theories that are consistent with this connected social learning have been developed. They include social-constructivism (Bruner, 1986; Vygotsky, 1978), the conversational framework (Laurillard, 2007), conversational learning theory (Sharples, Taylor, & Vavoula, 2007) and connectivism (Downes, 2005; Siemens, 2005).

Social constructivism is an extension of cognitive constructivism. The latter posits that knowledge is constructed by an individual through active interaction with learning materials. Prior knowledge is an important element in the internalisation of knowledge construction. In the social-constructivist model, knowledge creation results from engagement by the learner with the diversity of opinions and knowledge through interactions with peers, teachers, experts, friends, family and even strangers where culture and society will influence the learning. Hence, social-constructivism emphasises the collaborative nature of learning but retains a view of the individual as the one doing the learning.

Connectivism places less emphasis on the individual as the locus of learning and instead proposes that knowledge is distributed across social networks (across multiple individuals) and that learners connected to a learning community benefit from it, while also contributing information to it. Knowledge does not reside only in the mind of an individual nor is it found in only one location or device but distributed across multiple individuals and stored in a variety of digital formats. In these respects, connectivity and communicative capabilities enabled by mobile devices are two of mobile technology's more powerful frameworks to support social-constructivist and connectivist learning.

A third framework places the emphasis on the processes involved in learning with mobile devices. The learning theory that emphasises the communicative importance of mobile-enabled learning is the conversational learning theory proposed by Sharples, Taylor and Vavoula (2007). The theory conceives learning as dynamic conversational processes taking place across multiple contexts (technology with technology; technology with person; person with person) within systems where people and technology are in "continual flux" (p. 224). The learning theories of socio-constructivism and connectivism discussed above support this dynamic conversational process, but focus more on the knowledge and less on the processes through which knowledge is explored and created. Mobility allows the conversation-based learning to be situated, that is, as it normally occurs, and embedded within learners' real activity, context and culture (Brown, Collins, & Duguid, 1989; Lave & Wenger, 1990).

Laurillard's (2007) conversational framework is more descriptive of the processes involved, but she nevertheless asserts that mobile learning experiences, as interpreted through her framework can be richer, in situated, site-specific learning environments. Where learning is situated, it is motivating for students because they perceive the learning as authentic, relevant and are able to control and take ownership of the learning. Examples of conversations are instructions, discussions and feedback that could be text-based messages or messages in audio or video formats. The inherent ubiquitous characteristic of mobile technologies means that they could facilitate faster communication and foster collaborative experiences through their capacity to connect and respond to others in timely (anytime, anywhere) ways to share knowledge, ideas, opinions and experiences with others, hence optimising the individual's learning.

Mobile devices as pedagogical tools

Uses of mobile devices as pedagogical tools have been reported in numerous studies (e.g. Laurillard, 2007; Ng, 2011; Nicholas & Ng, 2009; Nykvist, 2012; Pegrum, Oakley, & Faulkner, 2013). The mobile-enabled social learning discussed above is itself a pedagogical activity where collaborative tasks designed by the educator are facilitated by mobile technology, either formally or informally. Mobile devices blur the boundaries between formal and informal learning, enabling continuity in learning between contexts. Pedagogically, mobile devices are tools that have the potential to support learning by providing a pervasive means of consuming information, communicating and sharing, investigating, capturing and analysing data, assessing, managing tasks and creating multimodal learning artefacts. The ability to support these learning activities is dependent on the capacities of the mobile devices, which as discussed above, are available with different features and capabilities. Some examples of the pedagogical use of mobile technology are as follows:

- consuming information by (i) accessing learning resources on the Internet such as *YouTube, Khan Academy, Scootle, Study Ladder* and *BBC* education websites, online interactive simulations, educational games and many more; (ii) listening to podcasts or viewing vodcasts created by the educator and/or other experts; (iii) learning with discipline specific apps e.g. *Chemistry Lab Suite* or *Today with History.*
- communicating and collaborating (i) via email, SMS or instant messaging e.g. *WhatsApp*; (ii) through videoconferencing using *Skype* or *Tango*; (iii) by placing comments on discussion forums in learning management systems e.g. *Edmodo, Moodle* or social networks e.g. *Facebook, Twitter, VoiceThread* or *Blogger*; and (iv) using interactive tools such as *Talkboard* for iOS devices, a collaborative whiteboard app where team members can sketch, brainstorm, create and doodle together in real time.
- creating, presenting and sharing (i) digital artefacts such as digital stories using digital camera and movie making apps e.g. *MovieMaker* or *iMovie*; (ii) concept maps using *Mobile Inspiration, SimpleMind* or *IdeaSketch;* (iii) tutorial style videos on showme.com; (iv) presentations on *Office Mobile* apps for Android-based phones or *Keynote* for iPhone/iPad or *Prezi* app for iPad; and (v) artefacts such as ePortfolios on apps like *Wordpress* and *Weebly* for both Android and iOS devices.
- undertaking formative assessment using *SurveyMonkey* and/or *PollEverywhere* for both Android and iOS devices; *Quiz Maker* for Android phone/tablet or *Quiz Creator* for iPhone/iPad.

There are also productivity apps, for example *OfficeSuite Pro* for Android smartphones and tablets that can be purchased at a low price. It has functionalities for documents, spreadsheets and presentations and it integrates with cloud storage services as well as offering pdf-markup commands.

EasilyDo is a virtual personal assistant app that is cost-free for both Android and iOS devices. The app proactively organises contacts, reminds the student of due dates and does other daily organisational things such as checking traffic and notifying the student about bad weather. Other useful pedagogical apps are GPS for collaborative location-based project work and voice recognition apps, such as *Dragon*, have mobile versions for most operating systems.

In a mobile learning context, the educator will need to consider the range of options exemplified above in order to provide the framework and structure for learning to take place. In a BYOD context, (s)he will need to be sufficiently digitally literate to understand the implications of multi-platforms with different features and make decisions about how best to employ the different features for learning. (S)he will need to understand that there are limitations with mobile learning. Limitations in processing power and display due to screen size convey the general perception that mobile devices are good for content consumption but are more laborious to use for content creation. However, it is necessary that the educator integrates both consumption and creation activities in order to engage students and foster the development of higher order thinking skills. (S)he will need to understand the delivery of content as small chunks and identify the purposes for which mobile devices are best used such as creating videos for flipped classroom teaching and learning. Likewise she will need to make decisions about how tasks set for students to create learning artefacts can be chunked, starting from small and building it up. In each case, this will involve choices. For example, students could snap a photo and use *Skitch* app (available for both Android and iOS smart devices) to annotate and share. Alternatively, students could snap and annotate a series of photos and build them into a slideshow or video using movie making apps for mobile devices.

As students externalise the thinking that goes on in their heads through external media, the educator needs to be actively exploring apps to identify the apps and the formats they offer that will enable students to externalise this thinking and the understanding of concepts learnt. The relationship between pedagogy and technology is not about one or the other, as is so often cited according to the slogan 'it's the pedagogy, not the technology'. There is tight interdependency of one on the other and the better the educator knows about the technology, the more creative and flexible (s)he will be in designing the pedagogy. Hence the more mobile digitally literate an educator is, the easier it will be to embrace mobile learning in the classroom in a sustainable way.

Mobile devices as personal tools for informal learning and conducting everyday activities

Mobile devices are highly personalised tools used for carrying out everyday activities. Young people live lives immersed in online environments. They access information from the Internet, interact with others online and conduct everyday activities such as checking bus times, doing their banking or negotiating a transaction (e.g. travel plans with a travel agency) online. Increasingly, young people are living their e-lives mediated through the mobile devices that they own. Many of these young people would have a degree of mobile digital literacy that they have developed informally and which has been motivated by friendship-driven and/or interest-driven online activities. While well equipped with a set of technical capabilities for these activities, many students would still need to be taught explicitly the cognitive and social aspects of using these devices and the applications that go with them, for both learning and day-to-day activities (Ng, 2012a).

Conceptualising mobile digital literacy

Mobile digital literacy is digital literacy associated with the use of mobile technology across all contexts of an individual's life. This includes the technical knowledge about using mobile technology, understanding of the formal and informal learning of curriculum and non-curriculum materials in mobile-mediated environments, the cognitive and social aspects of mobile-mediated activities and the day-to-day use of mobile technology in the individual's personal life. The term mobile digital literacy has been used by other researchers although the focus and contexts in which it has been used are different from those of this chapter. For instance, Houghton (2013) describes mobile digital literacy in terms of the use of different handheld devices, while Traxler (2012) discusses it in the more traditional sense of literacy, making an argument about the influence of mobile technology on context such as text and self as reader and writer. *Literacy* in this chapter is a pluralist construct that includes vital competencies such as the ability to identify, understand, interpret, create, communicate and compute texts (written, printed and digital) in the varying contexts of an individual's personal life and in the wider community (UNESCO, 2004). Within the UNESCO framework, the term has moved beyond the view of being equipped with the technical skills of reading, writing and being numerate. Texts in a digital world are no longer confined to just written text but include visuals (e.g. images, signs, videos), sound (e.g. lectures, conversations, music) and multimodal formatted artefacts (e.g. simulations, games and multimedia resources). In addition, Traxler (2012) asserts that mobile devices affect the way in which ideas, images and information are written, evaluated, produced, stored, distributed and read.

The mobile digital literacy framework, adapted from the mlearning literacy framework (Ng, 2013) has three dimensions: *technical* – relating to technical and functional skills, *cognitive* –relating to critical thinking, evaluative and multiliteracies capabilities and competencies, and *social-emotional* – relating to attitudes, social skills and safeguarding capabilities in mobile supported environments. As shown in Figure 5.1, within and between the dimensions are multiple literacies that need to be developed if we are to sustain mobile digitally literate individuals.

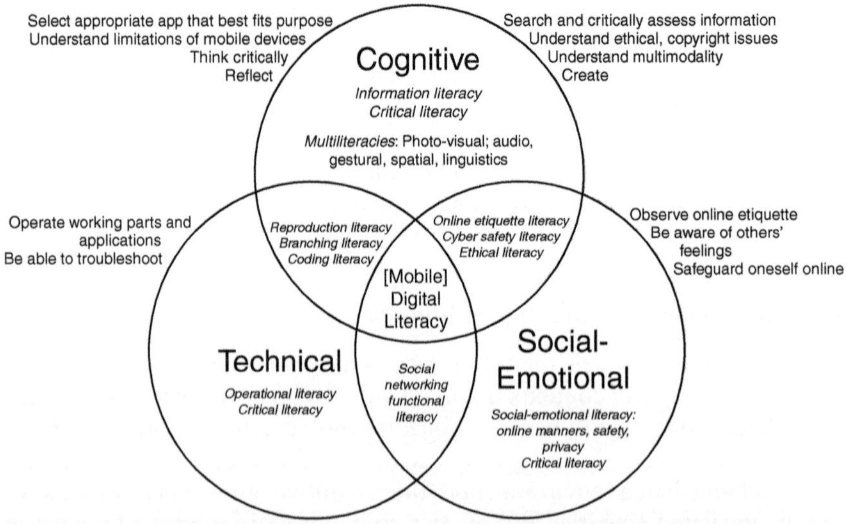

Figure 5.1 Mobile digital literacy framework, embracing the components in the three dimensions of cognitive, technical and social-emotional.

Technical dimension of mobile digital literacy

At the technical level, the mobile digitally literate individual is able to differentiate between the different mobile devices and how they differ technically and functionally. (S)he understands about the affordances and limitations of applications in mobile devices. Technical and functionally-related skills are developmental where ideally, the individual:

- is able to use input and the connecting peripheral devices such as touch screen keypads, navipad, stylus, half-QWERTY keyboard, earphones, Bluetooth headset, USB connector and printer;
- has knowledge of working parts for example, infra-red, Bluetooth, wireless, memory card and data synchronization; is able to configure basic device settings and use 'Settings' to control features such as sound and brightness; has general understanding of standard device keys and controls;
- is able to troubleshoot by reading manuals; accessing local 'Help' functions and/or web-based resources (e.g. *YouTube*) for assistance;
- understands file structures such as folders and directories; manages where files are located in the drive/memory card and knows how to access the drive via his/her computer (i.e. understanding about operating system compatibility);
- is able to manage data transfer, including understanding file sizes and spaces to hold files; is able to find, download, uninstall or delete and

installing applications that are compatible with the mobile device; understands data charge costs associated with downloading data; updates or changes user account information on the device and on the Internet; sends and retrieves attachments, opens them with appropriate applications including unzipping, uses infra-red and Bluetooth for file transfer; understands the limitations of word processing, spreadsheet and presentation tools; understands compatibility issues of operating systems, including the transferability of files between systems; understands 'cloud' computing and storage and the risks associated with it;
- knows about and is able to locate user interface elements, i.e. cues that define interactivity e.g. menu, sizing, scrolling, use of sliders, tabs and use of multiple windows for multitasking;
- is able to set up and use communication tools e.g. emails, web mail, VOIP, blogs, wikis, *Facebook*, *Google+*,*Twitter*, SMS, MMS;
- ensures that anti-virus software is regularly updated to avoid spam and viruses;
- is proficient in using mobile applications and resources for learning, creating and solving everyday problems. This includes being able to navigate through applications and Web-based resources and select the most appropriate features to accomplish specific tasks.

Developing proficient mobile technical and operational skills means that the individual needs to spend time exploring new applications and be provided with opportunities to practise, for example, through repeated use of mobile spreadsheets or presentation tools.

Cognitive dimension of mobile digital literacy

The crucial cognitive skills and competencies in mobile digital literacy are largely related to interactions with information on the Internet. A mobile digitally literate individual would use the Internet effectively for information gathering and synthesis. (S)he:

- is able to search, locate and assess web-based information; critique information by analysing and evaluating the digital content for accuracy, currency, reliability and level of difficulty;
- understands that many websites are not yet mobile friendly nor created for mobile devices and that a fair amount of scrolling (horizontally and vertically) will be needed to read information on the web;
- understands that mobile web pages could contain reduced versions of non-mobile web pages and that a balanced perspective of the article may not be complete;
- understands critical literacy and that people behind the scene producing the information have their own motives, hence is able to critically

evaluate whose voice is heard and whose is not in order to learn as neutrally as possible;
- understands the ethical and legal issues associated with the reproduction of digital resources, for example copyright and plagiarism policies; understands the terms and conditions of applications and resources well so that legal liability is avoided;
- is able to externalise understanding and synthesise new meanings using appropriate online or offline applications that will convey the meanings in the best format; is able to critically select the appropriate application for a task and knows that the capabilities of productivity applications could be reduced in mobile devices;
- understands about multiliteracies and is able to decode information from text, images, sound bites, videos, maps and models; similarly, (s) he should be able to demonstrate knowledge acquired and create digital artefacts using multimodal ways of representing meanings;
- has basic programming skills to author small apps or modify existing apps to serve a useful function.

Social-emotional dimension of mobile digital literacy

Millions of young people 'meet' online on a daily basis to chat, exchange ideas, communicate socially and collaborate on projects. Web 2.0 technologies such as *Wikispaces*, *Flickr*, *MySpace*, *Google+*, *Blogspot*, *Facebook* and *YouTube* are enabling individuals to contribute freely to networked communities for learning and/or for socializing. Web 2.0 is becoming a lifestyle for young people, who are accessing the Web to send emails, seek information, purchase goods, do online banking, chat online, post questions, contribute solutions, post photos and videos, download games, videos and music and write blogs and wikis to share ideas and opinions. These practices are enhanced through the ubiquitous characteristics of mobile technologies. Hence a mobile digitally literate individual would develop appropriate attitudes and behaviour in using mobile technologies for socialising and learning. (S)he:

- observes 'netiquette' by applying similar rules as in face-to-face communication such as respect and the careful use of appropriately abbreviated or SMS language to avoid misinterpretation and misunderstanding; is aware that SMS language can impact on formal language use in classroom situations; is able to interpret correctly instant and social network messages in terms of the tone of the message (e.g. use of bold letters and symbols/emoticons) and other underlying meanings;
- balances the amount of time spent on social networking sites and the number of text messages sent in order to manage costs, time to do other things and prevent social networking addiction;

- protects his/her own safety and privacy by protecting private information appropriately and not disclosing any more personal information than is necessary;
- recognises when (s)he is being threatened and knows how to deal with it, for example whether to ignore, report or respond to the threat; is able to interact strategically, ensuring safety by asking appropriate questions;
- is ethically literate and exercises moral judgement in using digital materials and participating in online activities;
- is aware of possible undesirable consequences for the actions that (s)he takes when trying out new ideas and taking on different personas in the virtual world; (s)he is aware of the fact that his/her online conversations, text or phone messages may be recorded permanently and what (s)he may think is private is in fact public.

Critical literacy

As shown in Figure 5.1, critical literacy is a constant element underpinning the development of each of the three dimensions of mobile digital literacy. Critical literacy is the ability to analyse and critique the relationships between texts, power and social justice, an ability that also involves challenging the values and beliefs that lie beneath the surface of texts. The term 'texts' in this context includes images, spoken and written words, music, songs, conversations, movies and other multimedia materials (Coffey, n.d.). As the amount of digital information that is readily available to us in different formats via mobile devices increases, developing skills to critically analyse the multitude of 'texts' to construct meanings from them is an important part of mobile digital literacy.

Freebody and Luke (1990) developed a repertoire of capabilities in their four-part resource model for critical literacy. These are: (i) code breaker (coding competence); (ii) meaning maker (semantic competence); (iii) text user (pragmatic competence); and (iv) text critic (critical competence). In taking the theory to the level of practice, Luke (2000, p. 454) identified the kinds of questions to ask for each component of the model as follows:

1 Coding practices: developing resources as a code breaker. How do I crack this text? What are its patterns and conventions? How do the sounds and marks relate, singly and in combination?
2 Text-meaning practices: developing resources as a text participant. How do the ideas presented in the text string together? What cultural resources can be brought to bear on the text? What are the cultural meanings and possible readings that can be constructed for this text?
3 Pragmatic practices: developing resources as a text user. How do the uses of this text shape its composition? What do I do with this text, here and now? What will others do with it? What are my options and alternatives?

4 Critical practices: developing resources as text analyst and critic. What kind of person, with what interests and values, could both read and write this naively and unproblematically? What is this text trying to do to me? In whose interests? Which positions, voices, and interests are at play? Which are silent or absent?

It is not within the scope of the chapter to discuss critical literacy in detail but the questions listed above are what mobile digitally literate individuals need to consider as they manage online information and resources. In particular, they need to be able to critique opinions on the Internet about contemporary topics such as global warming, genetically modified food and other social issues, and identify what information is missing and the authority of the opinions.

Critical literacy is not limited to critiquing online content and opinions. It also applies to social networking messages conveyed via the Internet or mobile network, where conversations and opinions are shaped by the worldviews of the people participating and may be biased in their expressions. At the technical level, critical literacy has a role in analysing how each type of technology provides the capacity to shape the texts and the choices made by the creator in assembling the digital material to represent his/her thinking.

In summarising, all interactions on the mobile device would require a degree of mobile digital literacy. The use of social media, the most popular use by young people, requires many careful considerations of the elements in the social-emotional and cognitive components of the mobile digital literacy framework, in particular the use of critical literacy in deciphering messages conveyed in dialogues, visuals, sounds and videos. In using the Internet, other questions that need to be addressed include "how do I find out about this person who wants to join my social network?", "who placed the *YouTube* video that I am now watching?", "what are appropriate things that I should say in the various blog sites that I am interested in?", "am I conveying the right meaning when I write this message or send this image/video?", "how private are the things that I write about online and will they have implications for my future life (e.g. career)?".

As an example of mobile digital literacy in practice, students using the *Good Reader* app on their iPads could be asked to peer review an assignment set by the educator using the annotation features available in the application. At the technical level, the student will need to know how to use the various features – when to tap, double tap, hold, draw bubbles, highlight, strike text, use sticky note etc. At the cognitive level, the student would need to be able to analyse the text critically and select the most appropriate feature(s) to annotate and provide feedback. For example, for short comments, the student could highlight and then write freeform with his/her finger a 1 or 2 word comment (such as "good" or "well argued"). For longer comments, (s)he could use the sticky note and type in the comments. At the socio-emotional level, the student will need to be sensitive in their choice of words to make the comments as well as the annotation styles (e.g. using bold or capital letters, crosses or smiley drawings) in order to convey the message in a constructive manner.

Some practical examples for developing mobile digital literacy in students

The best way to develop mobile digital literacy, for both school and higher education students, is to engage students in a variety of assignments or activities that make use of mobile technology. Activities should be integrated into the content of disciplines with discussions and problem solving tasks assigned to students to complete using the appropriate applications in their mobile devices. Developing mobile digital literacy in students should be within the context of the students' own mobile device. The better the student understands his/her own personal mobile device, the more effectively (s)he will use it for learning and everyday activities.

At the school level, it would be useful to dedicate several lessons in junior schools to develop students' mobile digital literacy explicitly, especially at BYOD schools. Some activities that could be included are as follows:

- Getting students to familiarise themselves with vocabulary that is associated with mobile technology and its use, for example, 3G/4G, server, service provider, etc.
- Getting students to know their own mobile device (phone, smartphone or tablet) by exploring and naming the physical features of their devices and ways they could be used, for example, how they would capture the content on the screen as an image and getting them to share the captured image with a fellow student.
- Getting students to articulate their phone plans. A 2013 study conducted in Australia indicated that the vast majority of the 1,129 adolescents who participated did not know what their phone plans entail (Ng & Nicholas, manuscript in preparation). Students could bring in their purchase agreements and analyse them. They could engage in some simple mathematics by solving problems on data usage and cost associated with texting, emailing, downloading movies, music and other resources, video Skyping, phone Skyping etc.
- Getting students to explore the file saving structure of their devices and to try saving or moving files to various storage spaces on their devices.
- Discussing the need for anti-virus apps in their mobile devices and what the advantages and issues are.
- Discussing and sharing mobile experiences associated with learning and day-to-day activities.
- Discussing mobile phone safety in terms of bodily health, for example, the types of waves emitted by mobile devices and what research says

about their impact on the brain and body; what the dangers are and how best to prevent health risks to the body.

- Exploring mobile-based messages recorded in their emails, SMS, instant message app and social network sites. Students could share their experiences and critique some messages that convey risks such as bullying; discuss strategies and develop skills to reply to potential conflicts and issues online. They could discuss their experiences in writing messages, including using abbreviations and images, on mobile devices and compare them with writing on laptops or desktops focusing on possible frustrations and how to deal with them. Students could also discuss and share with their teachers the types of writing activities that they would like to do on mobile devices.
- Engaging students in activities using multiple modes of representations to develop their visual, audio, spatial and multimedia literacies. For example, students could work in groups to analyse and elicit meanings from a series of images, illustrations, cartoons, videos and music pieces. They share their interpretations with peers and obtain feedback. Students could also undertake activities such as putting together imagery for a poem where they select a series of images or make use of other forms of visuals to illustrate the content of the poem. They will need to consider types of images, tone, colour and fonts. Another problem the students could solve is to create readability on a text-heavy page or on a small screen like the smartphone.
- Comparing various versions of online resources, for example, use of mobile versions of newspaper articles and compare the content with that from traditional laptop/desktop-based versions as well as with hard-copy versions. Students could analyse the layout, the display and the fullness of the reports.
- Reading information from mobile devices, analysing the information and sharing the individual's critique with the class. Students could focus on the affective aspect of screen-based reading on mobile devices, for example how fast and for how long can they read or does the small screen affect how much they can remember and what are the differences between screen-based reading on mobile phones, smartphones, tablets, laptops and desktops? They could discuss the types of texts that are suited for mobile devices for their learning.
- Engaging students in creating digital artefacts with their mobile devices. Each student could create a one-minute podcast on a specific topic/concept using the audio recording app and upload to the school learning management system or *iTunes* to share with his/her classmates. Similarly, the students could work individually using *MovieMaker* or *iMovie* on their mobile devices (or collaborate in small groups of two or three) to create short video podcasts (vodcasts) and upload them to the school's learning management system or *YouTube* for sharing with their peers.

- Engaging students in a series of problem-solving exercises based on mobile phone use, for example search on the Internet and discuss issues encountered using a mobile web browser. Another task could be students discussing the elements that need to be considered when viewing and buying a dress/shirt online via a mobile phone. Questions the students may consider are: Is the online store reliable? What are the indicators that show that they should purchase from this store? What is missing in the photos of clothes displayed? Are the features sufficiently clear to know that the clothing of interest is how the purchaser envisages it? How does the fabric feel? Which form of payment is safest? How much personal information should the purchaser (student) disclose? Is a money transaction carried out on a mobile device safe? How does the purchaser keep a record of what has been purchased? What is the best way to pay with a credit card without having to pay interest? What are the conditions of return? How will the item be delivered? What happens when the purchaser is not home?
- Getting each student to select an app (including a game app) and teach others how to use it by doing a presentation constructed on the student's own mobile device. Questions to consider are: What presentation app is compatible with the mobile device? What other means of presentation could be done without a productivity app equivalent to *PowerPoint* or *Keynote*?
- Teaching students how to create simple apps.

The list of examples is not exhaustive but would stimulate students to engage in mobile digital literacy development that embraces the cognitive, technical and social-emotional dimensions of the mobile digital literacy framework.

Conclusion

As ownership of personal mobile devices continues to rise, the development of mobile digital literacy is an inevitable part of education in today's contemporary society. Developing mobile digital literacy skills and competencies could improve attitudes toward the use of mobile technology for learning. By being skilled with using the features of his/her mobile device, the student's attention would not be distracted by the technical aspects of using the technology. (S)he would be able to work intuitively with the technology to focus the working memory on learning the material and undertaking the task at hand. In other words, being mobile digitally literate reduces the cognitive load imposed on the brain during learning with mobile devices. Apart from the technical skills, well developed cognitive and social-emotional skills for mobile learning will prepare students to access and use information on their mobile devices intelligently, ethically and safely. This is important as with the increased personal ownership and the ease in portability of mobile devices, the frequency of accessing Internet sites increases in a student's daily life.

There is a role for educators to foster the development of mobile digital literacy in all their students. However, careful planning is necessary to ensure the feasibility of tasks, including ensuring that the technical aspects (e.g. infrastructure, apps compatible with different platforms) of learning do not become a barrier. Educators embracing mobile learning programs need to ensure that students' mobile digital literacy is adequately developed in order to achieve better learning outcomes with mobile devices and sustain the learning across all contexts of their lives. An implication is that the educators themselves need to be similarly mobile digitally literate.

Notes

1 https://edutechdebate.org/meducation-initiatives/sms4learning-supporting-healthcare-providers-through-frontlinesmslearn/
2 http://www.rapidsms.org/case-studies/senegal-the-jokko-initiative/

References

ACARA. (n.d.). *Australian curriculum: Technologies*. Retrieved July 16, 2014 from http://www.australiancurriculum.edu.au/technologies/rationale-aims/technologies

AICTEC. (2009). *Digital education revolution implementation roadmap*. Retrieved March 28, 2014 from http://ehlt.flinders.edu.au/education/eduwiki/lib/exe/fetch.php?media=students10:research_docs:der_implementation_roadmap.pdf.

Brown, J. S., Collins, A., & Duguid, P. (1989). Situated cognition and the culture of learning. *Educational Researcher*, *18*(1), 32–42.

Bruner, J. (1986). *Actual minds, possible worlds*. Cambridge, MA: Harvard University Press.

Chan, T.W., Roschelle, J., Hsi, S., Kinshuk, Sharples, M., Brown, T., Patton, C., Cherniavsky, J., Pea, R., Norris, C., Soloway, E., Balacheff, N., Scardamalia, M., Dillenbourg, P., Looi, C.K., Milrad, M., & Hoope, U. (2006) One-to-one technology-enhanced learning: an opportunity for global research collaboration. *Research and Practice in Technology Enhanced Learning*, *1*(1), 3–29.

Coffey, H. (n.d.). *Critical literacy*. Retrieved May 12, 2014 from http://www.learnnc.org/lp/pages/4437?style=print.

del Carmen Valderrama Bahamóndez, E. & Schmidt, A. (2011). Mobile phones, developing countries and learning, In W. Ng (Ed.), *Mobile technologies and handheld devices for ubiquitous learning: Research and pedagogy* (pp. 120–137). Hershey, PA: IGI Global Publishing.

Department for Education. (2013). National curriculum in England: computing programmes of study. Retrieved March 23, 2015 from https://www.gov.uk/government/publications/national-curriculum-in-england-computing-programmes-of-study.

Downes, S. (2006). An introduction to connective knowledge. Retrieved April 13, 2014 from http://www.immagic.com/eLibrary/ARCHIVES/GENERAL/BLOGS/S051222D.pdf.

Ekanayake, S., & Wishart, J. (2011). Identifying the potential of mobile phone cameras in science teaching and learning: A case study undertaken in Sri Lanka. *International Journal of Mobile and Blended Learning (IJMBL)*, *3*(2), 16–30.

Freebody, P., & Luke, A. (1990). Literacies programs: Debates and demands in cultural context. *Prospect: Australian Journal of TESOL*, *5*(7), 7–16.

Helsper, E. J., & Eynon, R. (2010). Digital natives: where is the evidence? *British Educational Research Journal*, *36*(3), 503–520.

Houghton, R.S. (2013). Mobile digital literacy: Analysing the walking 'thought makers'. Retrieved May 20, 2014 from http://www.wcu.edu/ceap/houghton/edelcompeduc/ch1/handheldOS.html.

Ito, M., Horst, H., Bittanti, M., Boyd, D., Herr-Stephenson, B., Lange, P. G., Pascoe, C. J., Robinson, L., Baumer, S., Cody, R., Mahendran, D., Martínez, K., Perkel, D., Sims, C. & Tripp, L. (2008). *Living and learning with new media: Summary of findings from the Digital Youth Project*. Chicago, IL: John D. and Catherine T. MacArthur Foundation.

ITU [International Telecommunications Union]. (2013). *The world in 2013: ICT facts and figures*. Retrieved May 6, 2014 from http://www.itu.int/en/ITU-D/Statistics/Documents/facts/ICTFactsFigures2013-e.pdf

Kelly, T., & Minges, M. (2012). World Bank report 2012: Executive summary. *Information and communications for development 2012: Maximizing mobile*, (pp. 3–9). Washington, DC: World Bank. Retrieved May 5, 2014 from http://www.worldbank.org/ict/IC4D2012

Kennedy, G. E., Judd, T. S., Churchward, A., Gray, K., & Krause, K. L. (2008). First year students' experiences with technology: Are they really digital natives. *Australasian Journal of Educational Technology*, *24*(1), 108–122.

Keogh, K. (2011). Using mobile phones for teaching, learning and assessing Irish in Ireland: Processes, Benefits and Challenges. In W. Ng (Ed.), *Mobile technologies and handheld devices for ubiquitous learning: Research and pedagogy* (pp. 236–258). Hershey, PA: IGI Global Publishing.

Laurillard, D. (2007). Pedagogical forms for mobile learning. In N. Pachler (Ed.), *Mobile learning: Towards a research agenda* (pp. 153–175). London: WLE Centre, IoE.

Lave, J., & Wenger, E. (1990). *Situated learning: Legitimate peripheral participation*. Cambridge, UK: Cambridge University Press.

Littlejohn, A., Beetham, H., & McGill, L. (2012). Learning at the digital frontier: A review of digital literacies in theory and practice. *Journal of Computer Assisted Learning*, *28*(6), 547–556.

Low, L., & O'Connell, M. (2006, October). Learner-centric design of digital mobile learning. In *Proceedings of the OLT Conference* (pp. 71–82). Retrieved May 27, 2014 from http://www.academia.edu/download/30832501/learner-centric-design-of-digital-mlearning-low-oconnell-2007.pdf

Luke, A. (2000). Critical literacy in Australia: A matter of context and standpoint. *Journal of Adolescent and Adult Literacy*, *43*(5), 448–461.

Margaryan, A., Littlejohn, A., & Vojt, G. (2011). Are digital natives a myth or reality? University students' use of digital technologies. *Computers & Education*, *56*(2), 429–440.

Minges, M. (2012). World Bank report 2012: Overview. *Information and communications for development 2012: Maximizing mobile*, (pp. 11–30). Washington,

DC: World Bank. Retrieved May 5, 2014 from http://www.worldbank.org/ict/IC4D2012

Ng, W. (Ed). (2011). *Mobile technologies and handheld devices for ubiquitous learning: Research and pedagogy*. Hershey, PA: IGI Global Publishing.

Ng, W. (2012a). Can we teach digital natives digital literacy? *Computers and Education*, *59*(3), 1065–1078.

Ng, W. (2012b). *Empowering science literacy through digital literacy and multiliteracies*. New York: Nova Science Publishers.

Ng, W. (2013). Conceptualising mlearning literacy. *International Journal of Mobile and Blended Learning*, *5*(1), 1–20.

Ng, W. & Nicholas, H. (manuscript in preparation). Investigating Australian adolescent students' use of and attitudes toward mobile technology for learning.

Nicholas, H. & Ng, W. (2009). Ubiquitous learning and handhelds: An overview of theory and pedagogy. In P. Rogers, G. Berg, J. Boettcher, C. Howard, L. Justice, & K. Schenk (Eds). *Encyclopedia for online and distance learning*, Volume IV (pp. 2171–2176). Hershey, PA: IGI Global Publishers.

Nykvist, Shaun S. (2012). The trials and tribulations of a BYOD science classroom. In Yu, Shengquan (Ed.) *Proceedings of the 2nd International STEM in Education Conference*, Beijing Normal University, Beijing, China, pp. 331–334.

Odaka, K. (2013). *Use of smartphones in 14 Asian cities*. Tokyo, Japan: Hakuhodo Inc. Retrieved November 11, 2014 from http://www.hakuhodo.jp/pdf/2013/20130809_2.pdf

Ofcom (2013). *Children and parents: Media use and attitudes report*. Retrieved March 8, 2014 from http://stakeholders.ofcom.org.uk/binaries/research/media-literacy/october-2013/research07Oct2013.pdf

Pegrum, M., Oakley, G., & Faulkner, R. (2013). Schools going mobile: A study of the adoption of mobile handheld technologies in Western Australian independent schools. *Australasian Journal of Educational Technology*, *29*(1), 66–81.

Pew Internet Research. (2012). *Digital differences*. Retrieved March 8, 2014 from http://www.pewinternet.org/files/old-media//Files/Reports/2012/PIP_Digital_differences_041312.pdf

Rideout, V. (2013). *Zero to eight: Children's media use in America, 2013*. Retrieved March 10, 2014 from http://www.commonsensemedia.org/research/zero-to-eight-childrens-media-use-in-america-2013.

Saylor, R. (2011). *Passing the torch. 2010–2011 mobile learning report* (pp. 7–8). Abilene, TX: Abilene Christian University.

Sánchez, J., Salinas, A., Contreras, D., & Meyer, E. (2011). Does the new digital generation of learners exist? A qualitative study. *British Journal of Educational Technology*, *42*(4), 543–556.

Sharples, M., Taylor, J., Vavoula, G.N. (2007). A theory of learning for the mobile age. In R. Andrews & C. Haythornthwaite (Eds.), *The SAGE handbook of e-learning research* (pp. 221–247). London: Sage.

Siemens, G. (2005). Connectivism: A learning theory for the digital age. *International Journal of Instructional Technology & Distance Learning*, 2(1), 3–9.

Tally, B. (2007). Digital technology and the end of social studies education. *Theory & Research in Social Education*, *35*(2), 305–321.

Thornton, P., & Houser, C. (2005). Using mobile phones in English education in Japan. *Journal of Computer Assisted Learning*, *21*(3), 217–228.

Traxler, J. (2012). Context as text in mobile digital literacy: A European university perspective. In M. Specht, M. Sharples, & J. Multisilta (Eds). *Proceedings of the 11th International Conference on Mobile and Contextual Learning 2012*, (pp. 289–293). CEUR-WS.org.

Trucano, M. (2014). A 'mobile first' approach to educational technology. Retrieved May 26, 2014 from http://blogs.worldbank.org/edutech/mobile-first-edtech

UNESCO (2004). *The plurality of literacy and its implications for policies and programmes*. Paris: UNESCO.

University of Cambridge (2010). *Mobile communications for medicine*. Retrieved May 22, 2014 from http://www.cam.ac.uk/research/news/mobile-communications-for-medicine/

Vygotsky, L.S. (1978). *Mind in society: The development of higher psychological processes*. Cambridge, MA: Harvard University Press.

Weiser, M. (1991). The computer for the twenty-first century. *Scientific American*, September issue, 94–10.

6 Supporting sustainability and innovation of mobile learning in a UK higher education institution

Tim Goodchild and Andy Ramsden

Introduction

The landscape of higher education, and more specifically, healthcare education is rapidly changing. The challenge of delivering high quality, effective education in a time of rapid change is leading to increased pressures on universities and academics to be innovative and sustainable in their teaching and learning models. In part this is due to the cycle of emerging technologies, economic constraints, the increasing power and freedom of students and evolving pedagogical paradigms.

Goodchild and Chenery-Morris (2011) outlined the emergence of podcasts at a UK university (University Campus Suffolk (UCS)) between 2007 and 2009, with podcasts in the nursing and midwifery context including audio anatomy and physiology podcasts, clinical skills videos and interviews with service users. At the time these were innovations for course teams and students, and were supported by teaching and learning development grants. Five years later, in 2014, these resources are still in use, albeit in an evolved way. Podcasts are still used with students, but they are used alongside other resources such as *YouTube* videos, learning objects and e-books. It is questionable how far the use of podcasting as an example has percolated through the institution beyond the initial academic champions and module teams.

This chapter will explore the use of podcasts since their inception at UCS, to their current use alongside other multimedia solutions, and the introduction of more recent mobile learning technologies such as smartphones and tablets. It will also engage with an emerging debate around learning technologies and innovations as seen amongst academics. Technology enhanced learning and mobile learning as part of that, conveys the assumption that the technology will enhance teaching and learning for both academic and student, and there is clearly a growing use of emerging technologies within universities across the higher education sector. Within the nursing teams there is widespread use of simulation technologies, virtual learning environments, and media which is accessed on phones and tablets such as video and audio and social media, but it is also true that for nursing in particular, there is a focus on non-technological methods of teaching and learning. Dearnley, McClelland and Irving (2013), in a literature review of innovation in healthcare education, state clearly that "non-technological innovations were found predominantly within nursing" (more so than for medical or allied health professional education). The theme of trust will be explored as a method of sustaining mobile learning – trust between course teams who may be sceptical or wary of learning technologies, and those teams in higher education who support their use and integration. The necessity of personal engagement with academics and pedagogical effectiveness will be discussed in relation to sustainability and the question of whether technology does or can really enhance learning. This chapter will argue that there is disquiet among nursing academics over the plethora of learning and mobile technologies in a curriculum where care and compassion for people remain the main focus. The experience at UCS is illustrative of wider research findings and will be discussed alongside the differing viewpoints related to teaching practice and sustainability, with the conclusion that knowledge and support of the diversity of staff experiences, skills and motivations is critical to supporting and promoting sustainable, innovative and effective use of technology in healthcare education.

Podcasts and mobile learning at UCS: Explanation for the current pattern of use

Goodchild and Chenery-Morris (2011) described a number of effective uses of podcasting within their teaching, learning and assessment model and highlighted the positive impact it has had on the student learning experience. Their initiatives were funded through an institutional learning and teaching grant. The question from the institutional perspective is, to what degree have these developments and emerging good practice transferred to other course teams?

As expected, this is a difficult question to answer, for a number of reasons. Firstly, academics are increasingly taking ownership of the process of creating and distributing audio and/or video files; this makes monitoring more challenging. Secondly, the definition of podcasting has become broader over time, and now includes video-based media as well as purely audio. Consequently, this discussion may underestimate the actual use. The emerging pattern indicates the pilot was in the vanguard of developments at UCS, and it will have contributed towards the adoption of multimedia centric learning designs by other individuals and course teams.

In recent years the higher education sector, and in particular UCS, has witnessed an increase in the use of multimedia in teaching, learning and assessment. This is evident from different perspectives. The UCS Elevate Team (a central e-learning Team with an av/multimedia function) manage an institutional *YouTube* channel (www.youtube.com/user/ucselevate)

Table 6.1 Uploads to Elevate Team YouTube channel by year

	Number of files uploaded to Elevate Team YouTube channel	*Emphasis of content for learning*
2011	88	Small proportion
2012	276	The majority
2013	374	The vast majority
2014	443	The vast majority

which hosts video and audio learning objects on behalf of UCS course teams. The hosted material tends to include:

- role play, short video material to include in learning activities within the VLE;
- student presentations for formative feedback and/or summative assessment;
- capturing visiting lecturers' presentations.

The statistics illustrate the demand for the service from lecturers is continuing to grow. The service was launched in February 2011, and by the end of 2014 there had been well over 1,000 videos uploaded. Table 6.1 highlights the year-on-year rise. There has also been a shift in the focus of the materials, with materials for learning becoming the majority by 2012.

Of course, the use of *YouTube* as a service within society is now monumental with billions of video views per year, and it would be expected that this pattern would be mirrored within the sector as a whole. However, individual and departmental *YouTube* channels are also evident at UCS, implying a significant uptake in the use of multimedia in learning and teaching at UCS.

Secondly, a review of how staff use the Virtual Learning Environment (VLE) at UCS (Ramsden, 2014a), includes a snapshot of a random sample of available and active courses in 2012–13. The methodology involved identifying courses which contained audio/video files, and the embedding of third party content. A sample of 163 courses identified:

- 10% included uploaded audio files;
- 19% included uploaded video files;
- 11% included embedded third party materials.

These findings imply 10 to 15% of active courses within UCS's VLE within a given academic year will include integrated multimedia, which is either uploaded directly, or embedded third party content. On review, a typical use is illustrated by the Introduction to Social Research Skills Module [BMPPSY110], which:

1 includes third party material (audio and video) learning material;
2 provides scaffolding to students around assignments in terms of question interpretation;

3 provides audio, generic formative feedback from an assignment based on a lecturer's round table discussion.

Finally, insights from a number of data collection points on tablet technologies at UCS indicate a growing interest in using multimedia in teaching, learning and assessment (Elevate Team, 2013; Ramsden, 2012). There is increasing interest in using iPads to provide multimedia rich feedback. This includes the use of iPad applications (*iAnnotate*) to embed audio feedback within student scripts, and (*Explain Everything*) to create short and effective generic feedback videos around problem sets and solutions. A staff survey in 2013 on the use of iPads in the classroom received positive feedback from those academics who had utilised iPads with their teaching, primarily due to the ease of use of tablet technology and software. For example, feedback from three academics included the following:

- "iPad allows critiques, feedback and grading to be recorded and sent to students."
- "In the classroom, videoing presentations and discussions, photographing flip chart feedback and quick Internet searches."
- "In smaller group work the iPad has been given to students (during the session) to explore the app (anatomy) and piece theory together with the content they see."

All the above tasks (feedback and grading/videoing, photos and Internet/exploring multimedia anatomy) could be done simply with a range of other technologies such as phones, laptops and cameras, but it seems that the attractiveness of ease of use and integrated hardware and software of a tablet such as the iPad may mean that it was used by staff in this way. Of note also is that there is little information on the sustainability of these 'trials' of using the iPad in classroom teaching, and that all respondents had volunteered to take part in iPad trials, and were already engaged with technology enhanced learning in some form.

Whilst there is some adoption of multimedia across UCS, it must be noted that the use of multimedia (including podcasting) is not evident in the majority of taught modules across the institution. The evidence in general raises the questions around effective embedding and sustainability of learning technologies at the institutional level, and the transfer of ideas between staff. Historically, the pattern has been of champions or high-level users of technology-enhanced learning engaging in a meaningful way, and it is open to question whether the move to an ever greater variety of technological tools in teaching and learning is widespread amongst academics at UCS.

The Collis, Peters and Pals (2001) 4Es model is the starting point when developing a framework to understand and stimulate the adoption of a learning technology by an individual lecturer within their teaching, learning

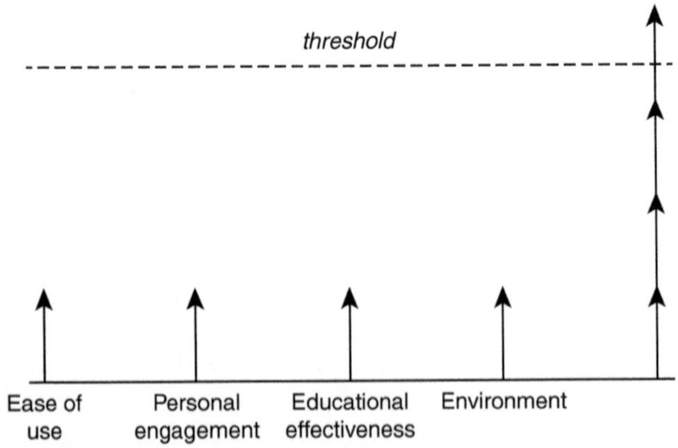

Figure 6.1 The 4Es model (revised).
Source: Adapted from Collis, Peters & Pals, 2001.

and assessment models. The 4E's model can be applied for predictive, comparative and interventionist approaches.

Figure 6.1 illustrates the revised 4Es Model. It predicts that the likelihood of an individual lecturer effectively embedding a learning technology within their teaching is a function of four broad factors:

1 environment (the institutional context);
2 education effectiveness (perceived or expected);
3 ease of use;
4 engagement (personal response to technology and change).

The model suggests if the combination of these four factors is above a perceived threshold the individual will adopt the technology within their learning and teaching. The model can be readily applied to understanding why podcasting is not more widespread within course teams.

Analysis suggests the two most important factors within the model are the environmental context and the level of personal engagement, where the key component of personal engagement is "I can picture myself trying to engage my colleagues or fellow students to try out a new [telematics] application for their learning related tasks". On reflection, since 2011, UCS has tended to place a greater emphasis on influencing the environmental context, as compared to the level of personal engagement of the individual lecturer. For example, since 2011, a number of institutional initiatives have included:

1 creating a central learning technology team with the aim to support staff using technologies within their teaching, learning and assessment;

2 improved and simplified learning technologies (software) and workflows;
3 creating a central multimedia production service, including the loan of equipment;
4 institutional (top down) initiatives: e.g. teaching development grants, annual teaching and learning days, and faculty teaching and learning committees.

However, there has been less coordination around empowering *the personal engagement factor*. This raises the question of whether this partially explains the slow adoption rate (institutionalisation) of well-established learning technologies, such as podcasting.

The level of personal engagement (through engaging with others) is associated with the spread of opinions through networks by opinion leaders and word of mouth communications. Kaiser et al. (2012) imply a few individual people within a network can exhibit significant influence. This implies social contacts, social interactions, and interpersonal communication are important influences on the adoption of new behaviours. The basis of word of mouth marketing is the general assumption people are talkers and we talk to satisfy distinct needs. Sernovitz (2012) suggests there are three reasons why people talk:

1 They like you and your services (such as an e-Learning Team, and innovation).
2 It makes them feel good to help and share.
3 It helps them feel connected as a group (a wider learning community).

Therefore, opinion leaders provide an essential peer education function within an organisation as they seek to engage colleagues to try out new approaches.

The importance of word of mouth communication and opinion leaders for adopting appropriate learning technologies can be visualised within Beetham's (2007) model of learning activity design. The model suggests practitioners consider several factors when designing an activity to engage and develop learners. One of these factors is other people (peers, tutors, facilitators, mentors and instructors). It would be reasonable to include the concept of "opinion leaders and influencers" within this factor.

So the speed of the diffusion and impact on personal engagement within the 4Es model is an outcome of how effective personal networks work within the academic community. Valente and Davis (1999) suggest the reason for low adoption is linked to the long time lag as the innovation percolates through networks before it reaches opinion leaders who are in the position to set the agenda for change. In other words, within our academic communities of practice at UCS the word of mouth communication around the innovations of Goodchild and Chenery-Morris (2011) have not been effective.

There is a growing pattern of use at UCS; however, this growth has been relatively slow. An apparent partial explanation is that central teams and individuals have not been prioritising the development of effective communication and dissemination strategies to get people talking. Looking towards the academic teams at UCS, it is noted that some teams seem more engaged with learning technologies than others, although whether this is because of particular people within those teams who are champions or resistant, or whether it is a discipline-specific issue is unknown. The podcasts which were introduced between 2007 and 2009, as highlighted by Goodchild and Chenery-Morris (2011), were used with nursing students, and continue to be used within nursing courses, with both authors residing within the Faculty of Health and Science at UCS. The nursing team at UCS utilise a range of learning technologies such as podcasting, simulation technology, and purely online modules via the institutional VLE, which make use of discussion boards and social networking. And yet, it is also apparent that there is a great deal of traditional classroom teaching on the nursing course which does not employ technologies and also some resistance noted from both students and academics to core technologies such as *PowerPoint*.

Despite *PowerPoint* being ubiquitous throughout education (Levasseur & Sawyer, 2006), phrases such as "Death by *PowerPoint*" are now so overused as to be considered cliché by most academics, and there seems to be little innovation or challenge to its accepted use. A research project at UCS which explored the use of *PowerPoint* as viewed by nursing academics and nursing students clearly found that in the main when it is used, it is done unreflectively, and as a teaching support tool for academics, rather than to promote learning, although some nurse academics avoided using *PowerPoint* for many sessions. Nursing students were also generally negative towards *PowerPoint* use in the course, and praised academics who tried to avoid its use. This simple technology, laced throughout courses in higher education, was actively avoided by some nurse academics, and is an example of the challenge that learning technologies may face within nurse education.

Nurse education and technology enhanced learning

Podcasts and mobile learning are among the many educational technologies which have permeated nursing and healthcare curricula over the years. When considering innovation and sustainability, it is important to take into account the context of mobile learning within healthcare education, and in particular for UCS, nursing education. There is an emerging debate within nurse education around the possible disjoint between technology-enhanced learning and academics educating students for a profession which focuses on values such as care, compassion and interpersonal skills.

Learning technologists (Coffait, 2012; Kitching & Wheeler, 2013; Spector, 2012) have for a long time promised enhancements and improvements in learning and teaching via technologies such as those involved in mobile learning. These asserted improvements are centred on claims that technology enhances teaching and learning (Open University, 2013); and despite the central role technology is afforded within this enhancement, the formal use of technologies, especially mobile learning, in higher education is best described as sporadic. For example, Ramsden (2014) in a survey of virtual learning environment use by academics at UCS found that the vast majority of academics (91%) were using the online module areas within the VLE as an administration tool, whilst less than 20% of modules on the VLE employed multimedia content suitable for mobile learning (see Table 6.2 below). This seemingly low use of tools such as podcasts within the VLE is indicative of a lack of engagement with learning technologies from many academics. Whilst there are a vast number of proponents of the benefits of learning technologies, there is a lack of systematic evidence in support of the enhancement aspect of learning technologies and the low engagement by academics, including mobile learning in the context of nursing education. This then begs the question of why does technology-enhanced learning persist and continue to evolve with new 'solutions' and who, if anybody, is promoting its use. Proponents of mobile learning also face sceptics within nursing education who have concerns about what matters in professional education, and what the best methods are for learning to be a professional nurse.

Nurse education has continually evolved through the last century alongside the many iterations of learning technology. In the 1990s nurse education moved into higher education in the UK with the implementation of 'Project 2000'. The Nursing and Midwifery Council is the professional body for UK nurses, and sets the education standards for all nursing courses. Today, UK nurse education is based upon the four domains of nursing competencies "professional values", "communication and interpersonal skills", "nursing practice and decision making" and "leadership, management and team working" (NMC, 2010). In the standards, technology is only mentioned twice in reference to patient care, and never in relation to education. However, there is a drive within nursing education to engage with technology, with the Prime Minister's Commission into nurse education (Keen, 2010) stating that "starting in their initial education, nurses need a better understanding of, and influence, over the development of technologies" and

Table 6.2 Sample of use of content on the UCS VLE 2014 (n=163 module areas)

Type of content	*Percentage of modules using that type*
Learning module	15%
Audio file	10%
Video file	19%
Images	11%
Embedded third party material	11%
External internet links	33%

that "nurse educators have a key role in terms of modelling the significance of ICT and e-learning skills in relation to nursing practice".

Over the last 30 years many technologies have emerged, and are often labelled as game changers in higher education. For example, Gleydura, Michelman and Wilson (1995), when discussing multimedia in nurse education during the heyday of the CD-ROM, introduce their discussion with the assertion that new developments in computer technology are changing the way we train professionals. Athappilly, Durben, and Woods (1994) argue that the CD-ROM is a tool to change the face of e-learning and that educators are convinced that "multimedia technology will help students to become more creative, more knowledgeable, and more curious about learning". Twenty years after this claim, it is hard to see how multimedia has changed the way we educate nurses, the face of e-learning or whether students are more knowledgeable, creative or curious. Arguably of greater impact for nurse education is the Internet. Commentators during the earlier years of the Internet lauded the impact it would have on nurse education; Lindeman (2000) believed that the era of the late 1990s was one of rapid change for nursing education, whilst Norris (1999) considered the Internet a tool for educators that "amplifies, extends, and even reorganizes human mental powers". However, the Internet cannot be taken alone as responsible for the evolution of nurse education – but rather can be viewed as a platform for tools. Technologies have for a long time appeared in the nursing classroom, for example Black and Watties-Daniels (2006) argued for "cutting-edge" technology to enhance learning in the nurse education classroom. However, Black and Watties-Daniels' definitions of a "smart classroom" or "web enhanced classroom" (projector, computer, DVD, Internet) focus on what technology is needed, rather than how this makes a classroom smart.

The majority of studies exploring the 'enhancement' aspect of technology-enhanced learning generally demonstrate little evidence of actual benefit above that of traditional methods of teaching (Lahti, Hatonen, & Valimaki. 2014), often with substantial barriers still present for student and academic (Worm, 2013). Wilkinson, Roberts and While (2013) published results of a longitudinal study into "nursing students' use of technology enhanced learning", and found that the technology employed during the course was a barrier to many. A literature review examining technology enhanced learning tools in nurse education (Petty, 2013) clearly indicates that efficacy of learning cannot be measured accurately, and that whilst there may be some benefits in the experience of learning (variability of teaching style) there is no demonstrable benefit to a students learning. It is stated in the conclusion that technology enhanced learning is "not embraced by everyone and engagement can be variable". Yet, despite this, the recommendation from Petty remains that "nurse education programmes should work towards developing [technology enhanced] tools for programmes of blended learning".

By using the phrase "technology enhanced" to describe the tools, Petty is assuming that the involvement of technology is an "enhancement". The issue of enhancement and effectiveness of technology-enhanced learning in higher education does have a wide research base, and as a backdrop to the findings of this chapter it is important to acknowledge this evidence base. It is clear from this evidence that technology-enhanced learning can have a positive impact on learning, although it can be difficult to quantify. For example, with classroom learning technologies, evidence suggests audience response systems (clickers) have positive effects on performance. Cavdar and Velasco's (2013) analysis suggests from a student perspective that clicker tasks aided them in understanding their lectures, the concepts discussed and other course material. These findings resonate with other evidence around perceived engagement and actual learning (Denker, 2013; Oigara & Keengwe, 2013; Welch, 2013). The positive, wider impact on learning is linked to various factors, including motivation as students compare their performance to peers, and self-improvement (Oswald and Rhoten, 2014). Although it is noted that the size of impact is strongly influenced by individual lectures and the alignment of the pedagogical model (Monk, Campbell, & Smala, 2013).

However, Keefe and Wharrad (2012) state that "many studies implementing e-learning are criticised for weak methodologies", and too often the factors of innovation and motivation by being part of a new experience are not explored. Keefe and Wharrad's (2012) study compared the learning achieved from a group of nurses who undertook standard training (low tech) with those who had access to both standard training and an e-learning tool (reusable learning object on pain assessment or management). Unsurprisingly, those who had access to an extra learning opportunity scored around 20% higher than those who did not, leading to the conclusion that the e-learning "enhanced" the knowledge of the participating students. However, surely a better comparison would be to compare a group of students who had access to the e-learning tool, with a group of students who had access to a low technology learning experience in addition to the standard training. The use of the word enhancement encourages the reader to consider that the technology being used is somehow better than non-technology based forms of teaching and learning.

The assumption is that technology and mobile learning offers enhancement, which results in the conclusion that non-technological methods of teaching and learning are 'incomplete' or need improving/enhancement. There is an assumption that more traditional methods of teaching preclude innovation, creativity or engagement, whilst it is clear in nurse education that "students continue to value traditional approaches" such as lectures and seminars, and that a continued focus on technological solutions will continue without "clear improvements in learning processes or outcomes for students" (Hall, 2009). Nurse and healthcare education has responded to evolutions in technology, and there have appeared to be many shifts in what learning technology is, but little or no movement in the paradigms surrounding education in response to this. Nurse education is considered by some to be dominated by educational theoretical perspectives derived from a broadly humanistic perspective (Quinn & Hughes, 2013), perhaps more so in nursing since the move in the 1990s from hospital based training, and more recently to a graduate only entry. Petit dit Dariel, Wharrad and Windle (2013) explored the factors influencing nursing academics in their choice to engage with technology, and found that there was a reluctance to engage with learning technologies, in part due to what they call "an ideological dilemma" between the changes in higher education, and the need to train nurses effectively. Odessa and Wharrad (2013), who investigated the factors influencing e-learning adoption in UK nurse education, pointed to the influence of a humanist educational narrative, with educators wanting to focus on human interaction in learning, rather than through a technologically mediated interaction. There was also a noted ambivalence toward the importance and influence of technology in learning and a perception that it has not had a significant impact in their own experiences.

Lahti, Hatonen and Valimaki (2014) have published what is the only literature review thus far on the effectiveness of e-learning in nurse education. They clearly state that there is a "lack of evidence" of the impact of technology in nurse education. They also conclude that "the results of this review showed that e-learning is not a superior learning method to traditional learning methods". They found no evidence that e-learning improves students' skills more than traditional methods, but some evidence that learning with technology is positively evaluated by students, and that learning with technology continues to have "potential". Selwyn (2014) posits that society suffers from a tendency to overlook our interactions with technology, sleepwalking through our encounters with technologies, and considering technology as somehow separate from the "messiness" of the everyday world. Technology is viewed as the zenith of society's achievements, and at the vanguard of progress in endless sectors such as medicine, archaeology, sport, environmental studies, and of course, healthcare and education. Mobile learning is becoming an integral part of the educational landscape, and as argued above is not a neutral entity which is the catalyst for educational transformation, but rather is something which is moulded and transformed by the actions of a wide variety of interests.

The focus of mobile learning literature is on how we might better harness the potential of mobile learning in education, and how we make the use of mobile learning in education more effective, rather than from a more critical perspective. Enhancements on account of technology continue to be put forward, and these improvements in teaching practice are part of a movement via a wide variety of blogs and social media use from academics and others involved in higher education. This steers discussion away from the political aspects of education and its inherent social nature, rather positioning technological developments as the most important cause of social change (Wajcman, 1994). Within universities we can look toward "Learning Technology" teams whose raison d'être is to find a way to make technology induced pedagogic improvements happen within their institution. They are an overt presence and can be found easily on the Internet via their own blogs and websites, and are an increasingly dominant group wielding power to influence and control the actions of academics and students within the institution.

Mobile learning has been heralded as inevitable in higher education for some years (New Media Consortium, 2014), perhaps because mobile technology is seen as part of our everyday life, mobile phones are omnipresent, and infrastructure and technologies such as mobile and wireless Internet access are becoming more pervasive. It is apparent via even a cursory glance through educational literature that there are many divergent views as to what mobile learning truly represents. "Mobile learning" has been through several iterations, initially with a technocentric approach where the focus was on the emerging technology of PDAs and then early mobile phones, with the term "handheld learning" emerging to finally represent this method of learning, where learning methods are attached to iterations of handheld technology (White et al., 2005). A more common expression of mobile learning today is that it refers to learning whilst mobile, or more specifically, learning that takes place away from the traditional sites of education such as a classroom or library where the focus is not on the devices, but rather on the mobility of the learners (Sharples, Taylor, & Vavoula, 2007).

However, it is clear for anyone within nurse education that there is a vast swathe of teaching that remains relatively untouched by learning technologies, especially mobile learning. There is also some debate as to whether students need or want technological innovations in higher education, with the HEA (2011) offering the conclusion that "there is no evidence of a pent-up demand amongst students for changes in pedagogy or of a demand for greater collaboration". This is in contrast to what is often reported in small-scale research involving technology, when students identify that the technology was a boon, aided their learning or enabled a better experience (Kirkwood & Price, 2014). Tellingly, the Higher Education Funding Council for England (HEFCE, 2011) also note that "not all staff are willing, or able, to engage with technology", which suggests that staff are being a hindrance or are incapable. The solution for this attitude or skill problem is to "increase support" for using technology or to "ensure they are sufficiently aware of technology". Again, it is a given that technology is necessary and will enhance teaching and learning. There is obvious commercial interest in the use of technologies in education. It is the 'for-profit' sector which is responsible for supplying hardware, software and content, the 'selling' of the need for technologies in education, and which also has influence in the formation of governmental and institutional education strategy – just head to an academic conference on education and learning technology to witness the marketisation of teaching practice. There are also managerial concerns with learning technology in today's economically lean times, and perhaps senior managers in HEIs are not as concerned with the transformation of pedagogy, but more focused on concerns of efficiency, cost-effectiveness, productiveness, student ratings and modernisation.

The Higher Education Academy promotes the development of university staff in the UK, and also the use of learning technologies via a programme titled 'Changing the Learning Landscape' which states that the workshops offer the chance "to explore how learning technologies can enhance their work in practical and creative ways". The HEA also include technologies as "core knowledge" within the UK Professional Standards Framework, asking academics to evidence their engagement with technology in respect of their teaching. There is a wealth of think tanks such as 'Blue Skies' in the UK and NMC Horizon Report (New Media Consortium, 2014) eulogising the use of technology in higher education. For example, the Blue Skies report (Coffait, 2012) states that learning "will become more personalised and achieve better outcomes than ever before". There is little or no critique offered in these reports, and yet nurse education sits within these frameworks and is overtly influenced by these organisations and their doctrines.

It could be argued that the potential vitality of technological solutions is lost in nurse education, when students look to the patient, the person, who sits in front of them for their practice – the art of nursing. Technology has reached such widespread adoption throughout society and healthcare that its apparent lack of effective penetration in nursing education is perhaps, more apparent than ever. Is it that nurse educators have technologies laced (arguably unreflectively) throughout their lives, that they look at technology in education and the solutions available such as those associated with mobile learning, and may not see congruence with their aims as a nurse educator – the messy, ambiguous, and context-sensitive process of becoming a nurse? One perspective in nursing education is that of humanistic educational theory which is developed from a wide range of educational theorists including Dewey, Maslow, Rogers and Knowles (McIntosh, 2011). Humanistic theory is concerned with human growth and self-actualisation (Braungart & Braungart, 2008) and that emotions and feelings are key elements in the educational process, with this emphasis on the affective aspects of humans being viewed as of equal importance to the cognitive and psychomotor domains. The humanistic perspective in nursing education encourages a patient-centred approach in care, with the main principles in this approach being the teacher-student interpersonal relationship and the classroom.

It is clear that the mastery of information is not the central project of the humanistic model, with Quinn and Hughes (2013) arguing that the application of the humanistic project in teaching practice is to enable strategies to arouse curiosity and to present problems and challenges, framed within a context of the "real world" of nursing, and managed in a classroom environment which is safe, whilst encouraging interaction, challenge and debate. Vitally, the role of the teacher in this context is that of facilitator, encouraging engagement and individuality via interpersonal skills. Despite a lack of clear evidence to support the utility of a humanist perspective in education (McIntosh, 2011), it is still prevalent in nurse education (Odessa & Wharrad 2013), and that for a profession which deals with the complexities and distress involved in being human, it has obvious merits. The importance of "technology" for learning in nurse education belies what many nursing academics and students may see as their values – equality, humanism, person-centred-learning, care and compassion –, which are in line with the values of the nursing profession, and which may bring them into conflict with purveyors of technology enhanced learning, who may look to equality in education, openness, cost-effectiveness, mobility and remote learning.

There is little evidence that cultural and educational practices have been transformed from where nurse education was 10 or 20 years ago, and whether mobile learning projects such as podcasting are valued by students and academics at the same level as classroom based teaching. The hegemony of technology-enhanced learning is a commonsense understanding, and whilst the value of enhancing learning and education may be a driver, it seems apparent that there are a raft of other ideological values, drivers and interests impinging on the use of technology in education. The discourse of technology in education is based in part on a beneficial fantasy that learning needs to be enhanced or will be enhanced by its use, and the question remains as to whether technology, and in particular mobile learning, has transformed the educational experience of students or is rather a supplement to a low-tech pedagogical approach.

Developing, sustaining and supporting mobile learning at UCS

Earlier in the chapter, the background of a rapidly changing landscape with emerging pressures on academics to be innovative and sustainable with their models was set out. The question we are faced with is "how can UCS develop, support and sustain innovative practice in mobile learning?" The answer will depend on scale. For instance, Collis et al.'s 4 Es model illustrates at the macro scale that there are a number of interlinked factors which need to be implemented by a number of teams and individuals across the institution. This holistic approach identifies the need to support the diversity of staff experiences, skills and motivations, such as those inherent to nurse education, which is critical to developing, supporting and promoting sustainable, innovative and acceptable use of technology in teaching and learning in healthcare education.

At the micro scale the discussion is how can a central based e-learning team encourage sustainable mobile learning. The issues around dissemination of ideas through a community of practitioners was previously identified

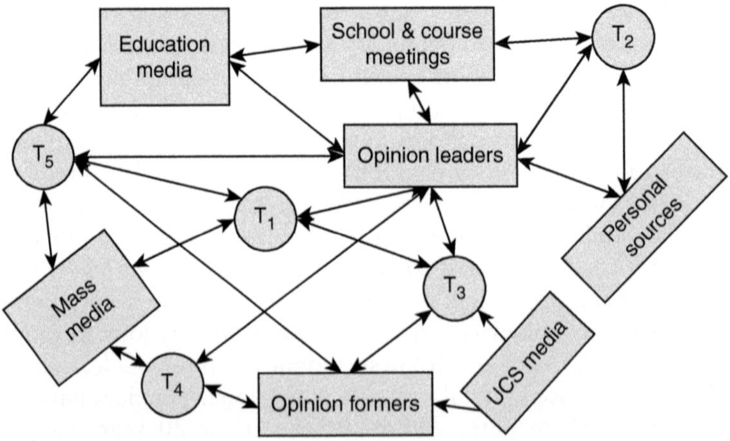

Figure 6.2 The role of opinion leaders and word of mouth communication in changing behaviour.
Source: Adapted from Fill, 2011.

as an inhibitor. To encourage communication, we need to visualise our dissemination activities within a connected network of practitioners (illustrated in Figure 6.1), where an academic (T) receives information from a number of sources, including opinion leaders, opinion formers, educational media, mass media as well as personal sources. They in turn will act as opinion leaders for others.

Interestingly, this network does not adhere to an institution's (hierarchical) organisational structure. If the institution's communication and dissemination strategy is primarily based on a top down organisational structure, with linear information channels, it is unlikely to foster a high level of personal engagement for a lecturer in a specific innovation. Consequently, the aim of the e-learning team is to help support and develop the awareness of opinion leaders, and make it easier for them to inform others. This is facilitated through the collection and sharing of information stories and experiences and helping to connect people with others. However, it is not simply disseminating the message. Sernovitz (2012) implies the person receiving the message must trust and respect the people/person delivering the message. This translates into four rules of word of mouth marketing (communication):

1 Be interesting. If delivery channels are boring you will never stimulate conversation.
2 Make it easy. Word of mouth is lazy, so help it along with simple messages which are easy to share.
3 Make people happy. When people like you they share you with their friends.

4 Earn trust and respect. Nobody talks positively about things they don't trust or like.

The adaption of this approach is evident with the Elevate Team (2014) at UCS, which has focused its online provision around collecting and sharing information on the effective use of learning technologies, and providing proportionally more resource to just in time support. This enables team members to continually engage with staff at all levels of the institution with the aim of helping staff make connections.

At the same time the factors of personal engagement, educational effectiveness and ease of use have been partially delivered through developing a course team using a cross-team development model around collaborating to develop online taster courses. The benefits include building trust, developing innovative solutions which are sustainable. Course teams can encounter new ideas in an easily digestible, discipline specific way, encountering a more integrated learning design, which uses an appropriate tool for that discipline. This is vital if we consider the emerging debates within nurse education as highlighted above. There seems to be a critical stance being taken, perhaps without knowledge of why that stance is being taken by some nurse academics. Whether it is from fear of new technologies, a lack of knowledge around the utility of technologies, a lack of support, or as argued a feeling of disquiet to the technological approach within the nursing discipline, mobile learning and podcasting have not become pervasive as teaching tools, and remain a minor solution in a teaching profession dominated by traditional teaching methods. A recognition from central support teams that there is pressure to acquiesce to technological solutions, and that some academics have the view that technology does not inherently enhance or transform learning can enable a greater sense of personal engagement within the central support methods from e-learning teams such as Elevate. A greater sense of trust can be built between academics and support teams, leading to the development and sustained use of acceptable and innovative mobile learning solutions.

References

Athappilly, K. K., Durben, C., & Woods, S. (1994). Cited in Reisman, S. (Ed.) (1994) *Multimedia computing* (103–124). Harrisburg, PA: IDEA Group Publishing.

Beetham, H. (2007). An approach to learning activity design. In Beetham, H., & Sharpe, R. (Eds.) *Rethinking pedagogy in a digital age*, London. Routledge, 1–10.

Black, C.D., & Watties-Daniels, A.D. (2006). Cutting edge technology to enhance nursing classroom instruction at Coppin State University. *ABNF Journal*, *17*(3), 103–106.

Brady, M., Seli, H., & Rosenthal, J. (2013). Metacognition and the influence of polling systems: How do clickers compare with low technology systems. *Educational Technology Research and Development*, *61*(6), 885–902.

Braungart, M.M., & Branngart, R.G. (2011). Applying learning theories to healthcare practice. In Bastabel, S., Gramet, P., Jacobs, K., & Sopczyk, D.L. (Eds.) (2011) *Health professional as educator*. Sudbury, MA: Jones & Bartlett Learning.

Cavdar, G., & Velasco, M. (2013). Teaching large classes with clickers: Results from a teaching experiment in comparative politics. *PS: Political Science & Politics*, *46*(4), 823–34.

Coffait, L. (Ed.) (2012). *Blue skies: New thinking about the future of higher education. UK 2012 edition*. Retrieved on March 12, 2015 from http://pearsonblueskies.com/wp-content/uploads/2012/09/Blue-Skies-UK-2012-FINAL.pdf

Collis, B., & Moonen, J. (2002). Flexible learning in a digital world. *Open Learning: The Journal of Open, Distance and e-Learning*, *17*(3), 217–230.

Collis, B., Peters, O., & Pals, N. (2001). A model for predicting the educational use of information and communication technologies. *Instructional Science*, *29*, 95–125.

Dearnley, C., McClelland, G.T., & Irving, D. (2013). *Innovation in teaching and learning in health higher education. Literature review*. Retrieved March 19, 2015 from: http://archive.councilofdeans.org.uk/Data/Sites/8/innovationinteachingandlearninginhealthhelitreview20130926.pdf

Denker, K.J. (2013). Student response systems and facilitating the large lecture basic communication course: Assessing engagement and learning. *Communication Teacher*, 27(1), 50–62.

Elevate Team. (2013). *TDAP: Elevate case studies: A new technology perspective (Tablet Technology)*, Retrieved on December 7, 2014 from https://docs.google.com/document/d/1SBbF4L1aUL-yYbZbkjeQ-OgjeX9scfQXoKtlbsT7btI/edit#heading=h.48e49h5m2p0b

Elevate Team. (2014). *About us*. Retrieved on December 7, 2014 from: http://ucselevate.blogspot.co.uk/p/about-us.html

Fill, C. (2011). *Essentials of marketing communications*. London: Prentice Hall

Gleydura, A.J., Michelman, J.E., & Wilson, C.N. (1995). Multimedia training in nurse education. *Computers in Nursing*. *13*(4). Retrieved on December 7, 2014 from http://digitalcommons.unf.edu/cgi/viewcontent.cgi?article=1003&context=bacc_facpub

Goodchild, T., & Chenery-Morris, S. (2011) Educational podcasts at University Campus Suffolk. In Ng, W. (Ed) (2011) *Mobile technologies and handheld devices for ubiquitous learning: Research and pedagogy*. USA. IGI Global Publishing.

Hall, W.A. (2009). Whither nursing education? Possibilities, panaceas and problems. *Nurse Education Today*, *29*, 268–275.

HEA. (2011). *Overcoming isolation in distance learning: Building a learning community through time and space*. Retrieved on December 7, 2014 from: http://www.heacademy.ac.uk/assets/cebe/documents/resources/workingpapers/WorkingPaper_18.pdf

HEFCE. (2011). *Collaborate to compete: Seizing the opportunity of online learning for UK higher education*. Retrieved on December 7, 2014 from http://www.hefce.ac.uk/media/hefce1/pubs/hefce/2011/1101/11_01.pdf

Kaiser, C., Krockel, J., & Bodendorf, F. (2012). Simulating the spread of opinions in online social networks when targeting opinion leaders. *Information Systems e-Bus Management*, *11*, 597–621.

Keefe, G., & Wharrad, H.J. (2012). Using e-learning to enhance nursing students' pain management education. *Nurse Education Today*. *32*(8), 66–72.

Keen, A. (2010). *Front line care: The Prime Minister's commission on the future of nursing and midwifery in England*. Retrieved on December 7, 2014 from: http://webarchive.nationalarchives.gov.uk/20100331110400/http://cnm.independent.gov.uk/wp-content/uploads/2010/03/front_line_care.pdf

Kirkwood, A., & Price, L. (2014). Technology-enhanced learning and teaching in higher education: what is 'enhanced' and how do we know? A critical literature review. *Learning, Media and Technology*, *39*(1), 6–36.

Kitching, L., & Wheeler, S. (2013). Playing games: Do games consoles have a positive impact on girls' learning outcomes and motivation? *European Journal of Open, Distance and E-Learning*, 2013/1.

Lahti, M., Hatonen, H., & Valimaki, M. (2014). Impact of e-learning on nurses' and student nurses knowledge, skills, and satisfaction: A systematic review and meta-analysis. *Journal of Nursing Studies*. *51*, 136–149.

Levasseur, D.G., & Sawyer, J.K. (2006). Pedagogy meets PowerPoint: A research review of the effect of computer-generated slides in the classroom. *The Review of Communication*, 6 (1–2), 101–123.

Lindeman, C.A. (2000). The future of nurse education. *Journal of Nursing Education*. *39*, 5–12.

McIntosh, A. (2011). Humanist learning theories. In McIntosh, A., Gidman, J., & Mason-Whitehead, E. *Key concepts in healthcare education*. London: Sage Publications.

Monk, S., Campbell, C., & Smala, S. (2013). Aligning pedagogy and technology: A case study using clickers in a first-year university education course. *International Journal of Pedagogies and Learning*, *8*(3), 229–241.

New Media Consortium. (2014). *Horizon project*. Retrieved on December 7, 2014 from: http://www.nmc.org/horizon-project

NMC. (2010). *Standards for pre registration nursing education*. Retrieved on December 7, 2014 from: http://standards.nmc-uk.org/PublishedDocuments/Standards%20for%20pre-registration%20nursing%20education%2016082010.pdf

Norris, P. (1999). Cited in: Chaffin, A., & Maddux, C. (2004). Internet teaching methods for use in baccalaureate nursing education. *CIN: Computers, Informatics, Nursing*, *22*(3), 132–142.

Odessa, D., & Wharrad, H. (2013). Exploring the underlying factors influencing e-learning adoption in nurse education. *Journal of Advanced Nursing*, *69*(6), 1289–1300.

Oigara, J., & Keengwe, J. (2013). Students' perceptions of clickers as an instructional tool to promote active learning. *Education and Information Technologies*, 18(1), 15–28.

Open University (2013). *Innovating pedagogy 2013*. Retrieved on December 7, 2014 from http://www.open.ac.uk/personalpages/mike.sharples/Reports/Innovating_Pedagogy_report_2013.pdf

Oswald, K.M., & Rhoten, S.E. (2014). Improving classroom clicker practices: Effects of incentives and feedback on retention. *North American Journal of Psychology*, *16*(1), 79–84.

Petit dit Dariel, O., Wharrad, H.J., & Windle, R. (2013). Exploring the underlying factors influencing e-learning. *Journal of Advanced Nursing*, *69*, 1289–1300.

Petty, J. (2013). Interactive, technology-enhanced self-regulated learning tools in healthcare education: A literature review. *Nurse Education Today*, *33*(1), 53–59.

Quinn, F.M., & Hughes, S.J. (2013). *Quinn's principles and practice of nurse education*. 6th edition. London: Cengage Learning.

Ramsden, A. (2012). *An iPad in the hands of a lecturer: An e-feedback pilot study at the University Campus Suffolk*. Retrieved on December 7, 2014 from http://bit.ly/HNZH0i

Ramsden, A. (2014a). *How are we using LearnUCS (VLE) at UCS?* Retrieved on December 7, 2014 from: http://andyramsden.wordpress.com/2014/03/12/how-are-we-using-learnucs-vle-at-ucs-first-sweep-of-large-data-set/

Ramsden, A. (2014b). Technology enhanced learning. *InsideUCS*, Issue 9 Spring 2014 (Digital) p. 10. Retrieved on December 7, 2014 from: http://www.ucs.ac.uk/About/RecentPublications/Inside-UCS/Issue-9/index.html

Selwyn, N. (2014). *Distrusting educational technology: Critical questions for changing times*. Abingdon: Routledge.

Sernovitz, A. (2012). *Word of mouth marketing: How smart companies get people talking*. Austin, TX: Greenleaf Book Group.

Sharples, M., Taylor, J., & Vavoula, G. (2007). Theory of learning for the mobile age. In Andrews, R., & Haythornthwaite, C. (Eds.) *The SAGE handbook of e-learning research*. London: Sage.

Spector, M.J. (2012). *Foundations of educational technology: Integrative approaches and interdisciplinary perspectives*. London: Routledge.

Valente, T., & Davis, R. (1999). Accelerating the diffusion of innovations using opinion leaders. *ANNALS, AAPSS*, *566*, 55–67.

Wajcman, J. (1994). Technological a/genders: Technology, culture and class. In Green, L., & Guinery, R. (Eds.) *Framing technology society, choice and change*, (pp. 3–14). St Leonards, NSW: Allen and Unwin.

Welch, S. (2013). Effectiveness of classroom response systems within an active learning environment. *The Journal of Nursing Education*, 52(11), 653–70.

White, A., Allen, P., Goodwin, L., Breckinridge, D., Dowell, J., & Garvy, R. (2005). Infusing PDA technology into nurse education. *Nurse Educator*, *30*(4), 150–154.

Wilkinson, A., Roberts, J., & While, A. (2013). Nursing students' use of technology enhanced learning: A longitudinal study. *Journal of Nursing Education and Practice*, *3*(5).

Worm, B.S. (2013). Learning from simple ebooks, online cases or classroom teaching when acquiring complex knowledge: A randomized controlled trial. *Respiratory Physiology and Pulmonology*, *8*(9). Retrieved on December 7, 2014 from http://www.plosone.org/article/info%3Adoi%2F10.1371%2Fjournal.pone.0073336

7 Sustaining mobile learning in inclusive environments

A universal design for learning approach

Therese M. Cumming, Iva Strnadová, Roselyn Dixon and Irina Verenikina

Introduction: Mobile learning within the UDL theoretical framework

Mobile technology is being adopted at a rapid pace by both schools and families, essentially being repurposed as assistive technology for students with disabilities (Gentry et al., 2010). This recent technology appears to have the potential to ameliorate many of the issues connected to the lack of sustainability of traditional forms of assistive technology implementation, which include: (a) stigmatisation; (b) cost; (c) maintenance; (d) sustainability; and (e) lack of alignment with the principles of Universal Design for Learning (Wehmeyer et al., 2008). As mobile technologies are being widely adopted as educational tools for all students, their potential importance as assistive technology for students with disabilities cannot be overstated. Many students with disabilities rely heavily on assistive technology to participate more fully in most environments in their lives. They use this technology for mobility (wheelchairs, electronic navigation systems), communication (Dynavox, PECS systems), social skills acquisition (video modelling), learning (computer software), literacy (screen readers, audiobooks), etc. (Merbler, Hadadian, & Ulman, 1999). Some students require several different devices; this can be impractical when moving from one environment to another. Through the use of a variety of personalised applications, mobile technology may possess the capability to provide many kinds of support with one device. For example, a screen reader application could be used to assist with literacy, a communication app with communication, and a movie app for social skills lessons.

Globally, students with disabilities have been included in the general education classroom in increasing numbers over the past few decades. The inclusion movement has left some general education teachers struggling to adapt their pedagogy to support an increasingly diverse student population. Global legislation such as the United States' *Individuals with Disabilities Education Improvement Act* (2004), Australia's *Disability Standards for Education* (2005), the United Kingdom's *Special Education Codes of*

Practice (2001), and the European Union's *Support for Children with Special Educational Needs* (2013) all provide frameworks to ensure that students with disability can access education on the same basis as other students. These policies require all education providers to make reasonable adjustments to ensure that students with disabilities are treated equally with their non-disabled peers. Often this is accomplished through the use of accommodations and modifications to the environment, the curriculum, teaching, and learning.

Accommodations and modifications are not always the optimal way to accomplish accessibility for all students. The environments and educational materials should be planned from the onset in such a way that they are accessible to all individuals (Edyburn, 2010). This underlines the importance of a teaching and learning framework that incorporates evidence-based practices that work for diverse populations of teachers and students from the outset. Accessibility is preferable over accommodation because it does not require extra effort in the form of time and resources, or moving to a special location, which is exclusionary. This type of built-in accessibility is an integral feature of Universal Design for Learning (UDL), which designs curricular materials and activities to be flexible in matching learner strengths and needs to assist students in reaching their learning goals (Gargiulo & Metcalf, 2010).

In order for assistive technology to be effective, it must be available across environments, and all individuals supporting the students must be well versed in the technology's use (Brown, 2011). If these conditions are not present, students with disabilities experience distinct disadvantage and difficulties with full participation in inclusive environments. Therefore, it is imperative that both theoretical and operational frameworks are available to guide the collaboration between school personnel and home in all areas of technology implementation. Universal Design for Learning (UDL) lends itself well as a theoretical framework to support using mobile technology as assistive technology to allow students with disabilities to more fully participate in inclusive environments. This framework is supported by the pedagogical approaches differentiated instruction and explicit instruction in inclusive learning and teaching settings.

Assistive technology

As found by Australian Institute of Health and Welfare (AIHW, 2008), students with disabilities experience many disadvantages educationally and otherwise. This is especially true for students with developmental disability, of whom 66% experience difficulties in learning, 41% fitting in socially, and 31% in communicating. These students are less likely to complete year 12 studies (19%) than their peers (49%). Only 9% of this population participates in any kind of post-school study, and as a group they are less likely to successfully transfer into the labour force (43%) than their peers (81%), which has a major impact on them individually as well as on the economy. If implemented properly, mobile technology has the potential to be a valuable assistive technology to improve both access and participation for students with disabilities.

Assistive technology devices are identified as: "Any item, piece of equipment or product system, whether acquired commercially off the shelf, modified, or customised, that is used to increase, maintain, or improve the functional capabilities of children with disabilities" (Authority 20 U.S.C., 2004). Multimedia and assistive technologies have a strong track record in the literature for producing positive learning outcomes for students with disabilities (Ayres & Langone, 2008; Banda, Dogoe, & Matuszny 2011; Braddock et al., 2004). Although various types of technology have been used in schools successfully with these students for communication, mobility, recreation, and teaching (Wehmeyer et al., 2008), there have been some significant challenges in providing the technology to the students. These have included cost, lack of portability, separate devices for each need, and complicated requirements for the devices' use and maintenance. Mobile technology may address many of these challenges, but research in this area is in its infancy (Cumming & Strnadová, 2012; Edyburn, 2013).

Educators globally are adopting mobile technologies such as smartphones and tablets for use in classrooms for everything from textbook replacement to assistive technology (Cumming & Strnadová, 2012). The New Media Centre touts mobile devices as technology that is likely to have a large impact on teaching, learning, and research globally (Johnson, Adams, & Haywood, 2011). Despite the devices' strong potential for individualising teaching, learning and communication, these are relatively new technologies, and the research evidence base to support their use as teaching and learning tools in inclusive environments for students with disabilities is limited. There is, however, promising early research on using these tools to improve outcomes for students with disabilities in the areas of communication (Campigotto, McEwen, & Denmans Epp, 2013; van der Meer et al., 2011); educational goal attainment (Ford & Rabe, 2011); basic academic skills (Fernández-López, Ródriguez-Fortíz, & Ródriguez-Almendros, 2013); and transitioning from one activity to another (Cihak et al., 2010). Early studies in the use of mobile technology as assistive technology for students with disabilities have been small and specialised and have not examined the broader area of effective implementation, which should include all of the students' environments and look ahead to the future, in order to promote sustainability of technology use.

Universal Design for Learning (UDL)

The inclusion movement dictates that students with disabilities receive their education in the general education classroom with their peers. This requires some adjustments to the planning and delivery of the curriculum.

The theoretical framework that best supports the delivery of the standard curriculum to all students is Universal Design for Learning (CAST, 2011a). Universal Design for Learning is an educational framework based on research in the learning sciences that guides the development of flexible learning environments (CAST, 2011a). It is defined as:

> a set of principles for curriculum development that give all individuals equal opportunities to learn. UDL provides a blueprint for creating instructional goals, methods, materials, and assessments that work for everyone--not a single, one-size-fits-all solution but rather flexible approaches that can be customized and adjusted for individual needs.

CAST (2011b) describes the three guiding principles of UDL: (a) provide multiple means of representation; (b) provide multiple means of engagement; and (c) provide multiple means of action and expression. These three principles, and the Universal Design for Learning model, are now widely accepted and recommended as part of different teaching models, including the popular Explicit Instruction (Hall, 2009) and Differentiated Instruction models (Tomlinson, 2008).

Classrooms that use mobile technology such as tablets and their associated applications (apps) are well-suited for all three of these principles. These devices and apps have the potential to allow more equitable access for students with disabilities in inclusive settings by allowing them the capability to individualise their supports in order to have every opportunity to learn, in the style that works best for them in both educational and community settings.

The concept of Universal Design began in the field of architecture, to ensure that buildings and other environments were built to be accessible to everyone from the start. Good design for people with disabilities benefits everyone, for example curb cuts and ramps were initiated for citizens with limited mobility, but have benefited people with strollers, wheeled luggage, and shopping carts. The same is true of UDL; providing multiple means of representation, engagement, and expression benefits all students, as it allows them to experience learning in the modalities they are strongest in.

Traditionally in the field of special education, accommodations and modifications to the environment, the curriculum, teaching, and learning are commonplace. This reduces students' opportunity for participation, however, as accommodation is usually dependent upon a request, which can take time. Often accommodation requires effort in the form of time and resources, or moving the student to a special location, which is exclusionary (Edyburn, 2010). To avoid this, UDL implementation involves planning the environments and materials in such a way that they are accessible to all individuals at the onset.

Research over the past two decades suggests that computer assisted instruction (CAI) is very effective for educating students, particularly those with disabilities. Palak, Walls, and Wells (2006) found that instructional technologies are powerful tools to achieve successful implementation of curriculum-based learning objectives to foster learning in regard to student engagement, collaboration and active learning. An emerging trend in the area of CAI is using mobile technology for educational purposes. Using a mobile device transforms CAI to mlearning, which increases functionality by allowing students to use computers as mind-tools from anywhere and at any time. It is especially useful in the inclusive UDL classroom, due to the support it offers in differentiating instruction for all types of learners.

Cobb (2010) suggests that differentiated instruction with Internet-based software is the best method of teaching urban school students, due to the great variability of their abilities and cultural backgrounds, and using a mobile device to access that software allows students to learn anywhere. Differentiating instruction with mobile technology allows for engaging students in different modalities, while varying the rate of instruction, complexity levels, and teaching strategies (Stanford, Crowe, & Flice, 2010). This assists teachers in working smarter, not harder, to meet the needs of their diverse learners.

It has been suggested that mlearning has the potential to close performance gaps for students with disabilities and minimise their classroom struggles (Seok et al., 2010). They further claim that technology integration provides the ability to adapt learning goals procedures, and materials for diverse learners, with or without disabilities, by increasing their access to the curriculum. Using a mobile device to access the curriculum is advantageous to students with disabilities for several reasons: (a) the devices have infinite patience; (b) students use multiple senses to interact with the device; (c) students can access the device and any material at their own pace; and (d) mobile devices can present information in ways that reduce cognitive load. Integrating mobile technology into instruction by using evidence-based practices and well-designed applications can level the playing field for students with learning disabilities (Seok et al., 2010). Mobile technology can therefore be used to differentiate instruction as per UDL guidelines. A brief description and examples of each follow.

Provide multiple means of representation

McGuire, Scott, and Shaw (2006) suggest that providing multiple means of representation enables more students to access a particular concept, and may enable students to engage more deeply with that concept. There are several ways that teachers can provide multiple means of representation. They can offer students different ways to perceive the information that is presented to them. This can be accomplished by offering alternatives for both visual and auditory information. Mobile technology can support teachers in this endeavour. Content can be presented electronically anytime anywhere in many different formats. Information should reflect the standards and essential skills required by the curriculum, but finding ways that students can choose studies within this curriculum will increase their interest, and potentially engage reluctant learners. CAST (2011b) recommends that students be provided with options for language, mathematical expressions and symbols. Using multimedia allows teachers to accomplish this in a variety of ways. Some of these include activities such as: webquests, videos, video lectures, and blogs.

Electronic textbooks have become quite popular and contain accessible features such as: (a) the ability to search for information within the text quickly and easily; (b) the ability to copy, paste, annotate, and highlight, which aids in note taking; (c) the ability to convert the text to an audio file, making it easier for auditory learners to access the material; (d) the ability to include links to related online material; and (e) the ability to incorporate a dictionary feature, which allows students to access the definition of unknown words simply by selecting a word. Mobile applications that focus on literacy, writing, mathematics, or specific content areas can also be employed to allow all students to receive the same content, at their individual levels, and at the pace each needs to learn effectively.

Provide multiple means of action and expression

When teachers provide multiple means of action and expression, they are providing students with a variety of ways of communicating and demonstrating what they have learned (Courey et al. 2013). It could be something as simple as allowing students to use a mobile device rather than traditional paper and pen to complete a writing assignment, or something more complicated, such as affording students the choice to take a test or write and perform a song to demonstrate what they have learned. In order to effectively uphold this principle of UDL, teachers need to provide students with choices, materials, and tools. Mobile technology provides many avenues with which teachers can differentiate the teaching and learning process. Teachers can use the technology to share information with students in multiple ways and students will have tools to enable them to collaborate away from school. For example, teachers can record their lectures using an audio recording application, and sync it with their lecture slides using an online presentation system such as *Prezi* or *Slideshare*, and students can access them on their mobile devices either online or through a presentation-viewing application such as *Prezi* or *Keynote*, at a time and in a place that is convenient for them, and review the information as necessary.

Communication is an essential component of learning, and mobile technologies can be used to support students during their face-to-face collaboration and communication with their peers. For students who have expressive language disabilities, electronic communication systems can be used to express their thoughts, needs, and opinions. Most of the currently available mobile communication applications are picture-based (i.e. *Proloquo2go, TaptoTalk,* and *AAC Speech Buddy*), although there are several keyboard-based applications (*AAC Text to Speech, Assistive Chat,* and *iMean* for example) available also (Farrell, 2013). This form of alternative communication is simpler, more portable, less expensive, and possibly most importantly, less stigmatising than previous methods and devices.

Provide multiple means of engagement

Courey et al. (2013) suggest that teachers can stimulate students' interest and motivation for learning by providing options that increase the relevance of instructional activities, such as following written directions to assemble a product. Providing students with real world activities and the tools to complete them encourages students to apply their knowledge and skills to solve problems that they find relevant. This may be the easiest place for teachers to differentiate, but if students are given a choice of activities, then teachers need to become comfortable with the level of activity that ensues when there are a variety of activities occurring simultaneously in the classroom.

Note taking can be a challenge for students with disabilities, particularly those with specific learning disabilities such as dyslexia (Gargiulo & Metcalf, 2010). Students who have difficulties with this task can use mobile technology as assistive technology. Writing and note-taking apps permit students to take notes by writing with a stylus, typing on a keyboard, or through verbal dictation. Some also allow for drawing, pdf annotation, and synchronisation with audio recording and cloud storage.

Drill and practice remains an evidence-based practice, allowing students to master material at their own pace, while providing them with feedback, and helping them to build confidence (Nunley, 2006). Electronic flashcard applications make it possible for students to practice without a partner. Content area apps and games are available in most subjects, but their usefulness is limited to specific topics. Teachers and students have repurposed generally focused apps such as word processors, calculators, spreadsheets and maps to suit their specific subject area lesson plans.

Product is concerned with how students demonstrate what they have learned. Assessment is an important part of a UDL classroom. Both formative (assessment that is ongoing during instruction) and summative (assessment that occurs at the end of the term or year) should be included in instructional planning. UDL prescribes multiple means of action and expression, so students may not be participating in traditional forms of assessment such as tests and quizzes. Teachers will need to determine how they will assess some of the non-traditional methods students can use to demonstrate what they have learned, such as presentations, portfolios, dramatic performances, and musical performances. Teachers can create rubrics to accomplish this. Rubrics are scoring systems that are used to evaluate a performance-based product (Gargiulo & Metcalf, 2010). Predetermined criteria are used to evaluate specific points of learning demonstrated by a project. Finson and Ormsbee (1998) found that rubrics are an accurate way of evaluating student achievement in the inclusive classroom.

Teachers can assess student learning by informally circulating throughout the room and asking students questions about their learning (Nunley, 2003). In addition to teacher-directed assessment, students can also demonstrate what they have learned in a variety of ways that suit their individual learning styles. Some examples of student work used to assess learning are: (a) paintings or sculptures; (b) plays or presentations; (c) songs; (d) diaries; and (e) traditional reports (Hall, Strangman, & Meyer, 2011). All of these can be created with and demonstrated via mobile technology, allowing students to work on their projects anywhere at any time, either individually or through collaboration with their peers. The role of teachers is to ensure that there are tasks to challenge each student, and that the expectations and requirements of which tasks are completed vary according to student ability. Students are provided with rubrics that provide them with the requirements necessary to demonstrate concept mastery.

Mobile technologies supporting students with disabilities in diverse educational settings

There are numerous claims related to the benefits of the use of digital technologies with students with disabilities. Specifically, the most prominent claims relate to their potential to teach and improve social skills (Cumming et al., 2008; Hopkins et al., 2011; Hourcade, Bullock-Rest, & Hansen, 2012,; Ozdemir et al., 2008; Verenikina et al., 2010; Xin & Sutman, 2011); facilitate communication (Charlop, Gilmore, & Chang, 2008; Shane et al., 2012; van der Meer et al., 2011); increase educational outcomes (Jowett, Moore, & Anderson, 2012; Kagohara et al., 2012; Mechling, Gast, & Krupa, 2007; Vedora & Stromer, 2007); reduce unwanted behaviours (Cihak, et al., 2010; Neely et al., 2013); and teach life skills (Shrestha, Anderson, & Moore, 2013).

Although still in their infancy, mobile technologies (particularly iPads, iPods, iPods Nano and iPods Touch) have gained considerable attention as innovative technologies, demonstrating their abilities to be effective pedagogical tools to support students with disabilities. This is evident in the growing body of literature advocating for its use within the special education setting (Douglas, Wojcik, & Thompson, 2011; Hourcade et al., 2012; Jowett, Moore, & Anderson, 2012; Newton & Dell, 2011). Similar to other digital technologies, these mobile technologies offer their users copious benefits, particularly in terms of educational outcomes, socialisation, communication and behaviour.

As a result, it is apparent that mobile technologies offer special educators pedagogical tools that have the potential to address many of the core deficits individuals with disabilities experience, predominantly in relation to social interaction and communication (Dixon & Verenikina, 2013). The various forms of mobile technology and its highly engaging and motivating nature offer an intriguing alternative to the high-tech, expensive tools that have previously dominated the market (Newton & Dell, 2011). Also, the inclusion of visual, auditory and kinaesthetic elements makes it a seamless option for students with disabilities, particularly as these elements draw on the motivators, strengths and interests that these students exhibit (Stafford, 2011). As a result, mobile technology has gained considerable attention as an innovative technology, demonstrating its ability to be an effective pedagogical tool to support students with disabilities.

Using mobile technologies to improve educational outcomes

The adoption of mobile technologies in special education has offered a new area for research since their introduction in 2010. The first articles describing how to use mobile technology in the classroom appeared in 2010 (Sailers, 2010); next came descriptions of how to evaluate apps for students with disabilities (Walker, 2011; Wilson, 2011). It wasn't until after this time that original studies appeared, therefore the research base is sparse. What is available is a range of studies in different educational settings from inclusive classrooms to segregated settings within schools. However, the number of studies and the number of participants is not extensive.

An influential systematic review found only 15 studies on mobile learning that met their strict inclusion criteria for empirical data (Kagohara et al., 2013). Most of the studies' participants had a diagnosis of autism spectrum disorder (ASD) and/or intellectual disabilities (mild to moderate). The majority of the studies used iPods or iPads. The results were very positive and suggested that these devices are effective educational supports for students with developmental disabilities. The results were particularly encouraging in the areas of academic, communication, leisure, employment, and transitioning skills. The authors suggested that more research is needed with larger numbers of participants and participants from different age groups and ability levels.

Overall, the existing studies have substantially demonstrated the capacity of mobile devices to effectively support the learning of numerous skills and strategies, essentially leading to improved educational outcomes (Jowett et al., 2012; Kagohara et al., 2012; Price, 2011). This can be seen in a recent study that employed a case study approach to explore the effectiveness of using an iPad-based video-modelling package to teach numeracy skills to a child with ASD (Jowett et al., 2012). The study returned highly positive results which may be attributed to the ability to personalise the approach by incorporating the child's interests (*Angry Birds*), using voice-overs, using the child's name, and shooting the video from a point-of-view perspective. These were valuable strategies that resulted in gains connected with motivation, enthusiasm and engagement (Jowett et al., 2012).

In another study, Kagohara et al. (2012) used an instructional video on an iPad with two participants with ADHD and Asperger Syndrome in an inclusive classroom to teach students how to use the spell checker on their computer. The results were very positive and suggested that iPad-based video modelling was very effective for a spelling intervention. Similarly, video self-modelling (VSM) with the iPad was used in a multiple-baseline design to teach functional mathematics skills to four adolescent males with ASD (Burton et al., 2013). Results indicated a strong relationship between using VSM and the ability for the participants to accurately answer the maths problems across three replications.

Further adding to the evidence that iPads are beneficial for teaching academic skills to students with disabilities is a study that was conducted to measure the effectiveness of using iPads as e-readers for students with ASD who exhibited significant deficits in reading comprehension (Price, 2011). The study involved three groups of ten students, with data collected over a one-year period. Findings indicated strong statistical evidence to suggest that interactive e-books (using iPads) dramatically improved reading comprehension. This was evident as all but one participant demonstrated sizable improvements in reading comprehension when using the iPad. These positive results were further verified due to the absence of any decrease in information acquisition arising from participants' use of the iPad (Price, 2011). In a Spanish study, Fernández-López, Ródriguez-Fortíz, and Ródriguez-Almendros (2013) designed an app for students with special educational needs. The app was fully customisable and included activities involving exploration, association, sorting and puzzles. The results of their pre-experimental pre-post-test design indicated that the app allowed some students to perform activities that were previously inaccessible to them, and that using the app had positive effects on the students' attainment of basic skills. The students showed improvements in language, mathematics, environmental awareness, autonomy, and social skills. Students also enjoyed using the app and were more motivated in their learning. The researchers surmised that mobile devices have the potential to assist people with intellectual disabilities to improve their quality of life and independence.

Gentry et al. (2010) examined the use of PDAs as task management tools for 22 high school students with autism. Occupational therapists taught the students to use the PDAs to manage tasks, and after eight weeks the students demonstrated significant improvement on an occupational performance measure, had learned to use the devices' reminder alarms and could independently program the devices. In addition, all of the students reported high degrees of satisfaction with using the devices, and felt that they improved their independence in performing functional activities, which in turn, improved their satisfaction with everyday life.

A study conducted by Ford and Rabe (2011) investigated a parent and teacher collaboration model of teaching a 20 year-old student with intellectual disabilities to use an iPhone as assistive technology to support him in meeting his academic and job site goals. They found that substituting mobile technology for his traditional assistive technology (Wolf, Dynavox, and other devices) improved his time management, enriched his connection with family and friends, empowered him and increased his self-determination. Gentry et al. (2012) also explored using mobile technology for task management, but with adults with autism. The participants successfully learned to use iPods as PDAs to improve their task completion in the workplace. The results of these two studies support the efficacy of portable devices as learning and cognitive aids for this population.

The combined results of these studies illustrate the considerable potential of mobile technologies to be used as a tool to support the improvement of educational outcomes and provide a justification for their sustained use for students with developmental disabilities.

Using mobile technologies to teach social skills and facilitate communication

A number of studies have employed the iPod Touch or iPad to teach both social skills and communication skills. A recent study involved five participants with ASD, multiple disabilities, or intellectual disability and explored the potential of the iPad to be used as a viable communication device for making requests when compared to a non-electronic AAC system involving graphic symbols (Flores et al., 2012). Conclusions drawn from this study indicated that the iPad did not detract from students' communication, but rather, behaviours related to communication either increased when using the iPad or were consistent with the picture-based system. The iPad did display strengths in terms of accessibility and less time required in set up compared to the picture system, which appealed to the instructors involved in the study. Achmadi (2010) employed a multiple-baseline across subjects design to investigate the use of an iPod as a communication device. The participants were taught to use a speech-generating application to make multi-step requests. Results were positive and suggested that all students could successfully learn to use an iPod for expressive communication. A comparable study was conducted by Kagohara et al. (2012) in a more segregated educational setting and investigated the prospective use of the iPad and iPod Touch to teach two adolescent students with ASD to name pictures using *Proloquo2Go* software as a speech generating device (SGD). The findings are highly favourable of the combined use of *Proloquo2Go* software and iPads and iPod Touches as a SGD.

Another study by van der Meer et al. (2011) aimed to teach three people with developmental disabilities to request toys using an iPod Touch. A multiple probe across participants design was employed, and the results suggested that two of the three participants were successful in learning to use the device to communicate. Van der Meer et al. (2011) cited several advantages to using mobile technology for the purpose of assistive technology to communicate: the devices are relatively inexpensive when compared to other AAC systems, students with developmental disabilities can learn how to use them via systemic instructional procedures, and they believed that the speech output from the iPods facilitated natural speech production. They also recommended that parents and teachers be trained to use, maintain, and troubleshoot the hardware and software and that others that communicate with the students be supported in using artificial speech. These studies all demonstrated that well-established instructional procedures were successful in teaching students with developmental disabilities to use an iPod Touch or iPad for communicative purposes. Other studies that focused on the communication domain for students with developmental disabilities (Kagohara, 2010; Kagohara et al., 2012; van der Meer et al., 2012a; van der Meer et al., 2012b) used similar conceptual procedural approaches and yielded similar results.

In a Canadian study, which was conducted in an inclusive regular high school, Campigotto et al. (2013) investigated the impact of the iPad-based intervention in a special education classroom for adolescents with a range of disabilities that included intellectual impairment, Down syndrome, ASD, and learning disabilities. The researchers concluded that the multimodal features of the device allowed for individual differentiation for this cohort of students, increased their perceptions of academic success and fostered an increase in self-confidence.

Using multi-touch tablets to enhance social skills among students with ASD was investigated within a study conducted by Hourcade et al. (2012). Within this study, the researchers worked with 26 students with ASD, their teachers and other stakeholders to develop a set of activities based on apps that run on multi-touch tablets such as the iPad. Results were extremely encouraging, with evidence to support the notion that activities using apps on the multi-touch tablet led to pro-social behaviours, essentially enabling students to engage in and enjoy social activities, as well as develop and express appropriate social skills (Hourcade et al., 2012).

A study by Hammond et al. (2010) taught three adolescent participants with intellectual disabilities how to engage in leisure skills, such as watching a movie using a third generation iPod Nano. Two out of the three participants successfully achieved the goals. In a further study related to leisure (Kagohara et al., 2011), three students with severe intellectual disabilities were successfully trained to independently operate an iPod Touch to watch movies. This study was also unique in that the instructional videos were delivered via the iPod Touch.

Video modelling is an accepted practice for teaching students with disabilities a variety of tasks and social skills. In a study by Cihak et al. (2010), video modelling delivered via an iPod was used as an effective tool to improve the independent transitioning of four elementary school students with autism in inclusive settings. Students increased their mean percentage of independent transitions from 7% to 77%. Teachers and students both reported satisfaction with the devices, particularly with their portability. Banda, Dogoe, and Matuszny (2011) reviewed 18 studies which examined video prompting for individuals with developmental disabilities. Two of those studies used mobile technology to present the videos, and the interventions were found to be effective for promoting the generalisation and maintenance of skills. Overall, the studies showed high social validity, with both parents and students having positive feelings towards the technology.

Using mobile technologies to reduce disruptive and challenging behaviours

Relatively few studies have examined the potential for the mobile technologies to address or reduce challenging and disruptive behaviours. Neely et al. (2013) conducted a study that compared the delivery of academic instruction using the iPad with traditional materials such as pen and paper, observing how this influenced the occurrence of challenging behaviours and academic engagement behaviour for students with ASD. It was found that both participants exhibited lower levels of challenging behaviour and higher levels of engagement when using the iPad, and higher levels of challenging behaviour and lower levels of engagement when using traditional materials. Conclusions from this study proposed that mobile technology offers great potential as a means of instructional delivery, with an ability to reduce unwanted and disruptive behaviours, whilst fostering engagement (Neely et al., 2013).

Changes to and transitions within the environment often present problems for children with developmental disabilities and can potentially result in behavioural issues (Cihak et al., 2010). The previously mentioned Cihak et al. (2010) study evaluated the effectiveness of video-modelling (VM) using handheld devices such as the iPad or iPod Touch to aid students in transitioning between classes and activities throughout the day (Cihak et al., 2010). Their findings revealed that not only was VM effective in facilitating independent transitions but also in decreasing inappropriate behaviours to zero during times of transitions, with skills still evident nine weeks after the intervention. These highly positive results offer an authentication for the sustained use of iPods and iPads to inhibit problematic behaviours during times of change and transition.

Overall it is evident that mobile technologies possess unique qualities and features advocating for them to be used by educators as a pedagogical tool to support students with disabilities. Such technologies have a proven ability to facilitate and increase communication, teach social skills, improve educational outcomes and reduce disruptive and unwanted behaviours.

Although research into the use of mobile technology to assist students with disabilities is in its infancy, early results are promising. Much of this research has been conducted in the areas of communication, transition, and engagement, and has been undertaken with students with emotional/behavioural disorders, autism and developmental disabilities. Through the use of applications that allow students to communicate, manipulate text, provide virtual experiences, self-monitor behaviour, and experience *hands-on* learning, mobile technology has the potential to support students with disabilities in participating more fully in inclusive classrooms. Lastly, mobile devices can also provide an important link between school and home by enabling students to access materials anywhere at any time.

Features that promote sustainability

It is obvious from the studies discussed above that mobile technologies (iPod, iPod Nano, iPod Touch, iPads, etc.) are very useful technologies for a range of students with disabilities and can be implemented in a variety of ways to achieve a variety of purposes. The literature cited specifically discusses the areas of academic, communication social and leisure skills and challenging behaviour. This wide range of praxis is indicative of the countless potential areas where these technologies could be included in a range of educational programs. The research base to date is not substantial, but the existing literature is nearly uniformly positive, although it has to be acknowledged that the empirical evidence in the inclusive setting for students with disabilities is sparse. Further research is needed to address the use of mobile technologies in a wider range of settings and with a wider range of students with disabilities, particularly those who are in inclusive environments.

The limited empirical data to date consistently support teachers' anecdotal reports that as tools for delivering educational programs for students with disabilities, mobile technologies have many advantages over other types of assistive technology. Specifically, they are pervasive, so they are very socially acceptable in most environments. They are relatively inexpensive, user friendly and operate intuitively. They are easy to individualise through the large variety of applications that are available. They are also less stigmatising than other devices and could facilitate inclusion in many classrooms. Lastly, since the technology is ubiquitous, most teachers are familiar with at least the basics of operating the devices. This promotes sustainability, although teachers also need to learn how to effectively integrate the technology into their pedagogy.

Barriers to the sustainability of assistive and mobile technologies

There have been several challenges to the use and sustainability of assistive technology, including stigmatisation, cost, mobility of devices, and difficulty in learning how to use the devices on the part of the students, teachers, and parents (Ford & Rabe, 2011; Jeffs, Behrmann, & Bannan-Ritland, 2006). Although mobile technology has the potential to overcome these challenges and offers individuals an assistive technology device that is portable, versatile, easy to use (even for young children), and fully customisable (Cumming & Strnadová, 2012; Cumming, Draper Rodriguez, & Strnadová, 2013), it is not without its own barriers. There has been limited research in the area examining mobile technology as a means of improving access and participation for students with developmental disabilities, and none that explores the challenges and successes of implementing the devices across environments. One of the anticipated barriers to sustainability is poorly designed apps. Parents or software developers design many of the educational applications for mobile devices, without consultation with education experts. Many applications that would be helpful specifically for students with disabilities never get designed, or if they do, do not reach the consumer market due to the lack of financial feasibility of catering to such a small percentage of the population (Braddock et al., 2004).

An issue related to this is one of teacher awareness of what applications exist and how to repurpose common applications to assist students with disabilities in inclusive settings. This barrier appears insurmountable, with over one million apps available and over 1,000 new apps entering the market each day (Koetsier, 2013). In order for mlearning to be sustainable, the devices and the applications must work across multiple environments (Braddock et al., 2004). This is especially important for apps that are used to assist students with daily living, environmental control, vision, hearing, recreation, mobility, reading, learning, studying, mathematics, motor aspects and composition of writing, communication and computer access. Teachers and parents must have some training to assist them in using and teaching students to use the devices, as well as enable them to recognise helpful apps (Cumming, Draper Rodriguez, & Strnadová, 2013).

Cost is another barrier that needs to be addressed. Although mobile technology is less costly and more prevalent than other forms of educational and assistive technology, the costs involved in widespread mobile technology adoption are still beyond the means of some students and schools. Added to that are the costs involved in upgrading the technology every few years, as it becomes obsolete. Although most mobile apps cost between $1 and $5, others, such as those that act as communication devices, at over $100, can be quite costly.

As with any form of educational technology, in order for implementation to be effective, teachers need to possess the knowledge necessary to incorporate mobile technology into their pedagogy. A large body of research in general education highlights that instructional technology integration has not been achieved, therefore it is imperative for teachers to become capable of integrating mobile technologies into their classroom pedagogy if students with disabilities are to benefit from the advantages which have been found in the empirical literature. Teachers' successful integration of technology is frequently hindered by a variety of mostly 'teacher-related' factors. The most frequently cited teacher-related factors that hinder technology integration encompass: (a) a lack of professional learning; (b) limited access to resources; (c) teachers' knowledge and skills; and (d) teachers' beliefs and attitudes (Ertmer et al., 2012). Most of these challenges can be met through professional development.

Sustaining mobile technologies through professional learning

A lack of, or ineffective professional learning programs are frequently cited as the main reasons for the lack of integration of technology (Balanskat et al., 2006; Beggs, 2000; Bingimlas, 2009; Schoepp, 2005). This is apparent as a worldwide survey found that the major obstacles that impeded teachers' technology integration specifically related to a lack of knowledge and skills, stemming from insufficient training opportunities (Pelgrum, 2001). An and Reiguluth (2011) found that technology training programs to this point in time have focused on attaining the skills needed to use the technology and had not emphasised the critical relationship between technology and pedagogy. They suggested that teachers need to be shown how to integrate technology into the curriculum and they need to do this in authentic teaching and learning contexts.

As both Bandura (1997) and Mumtaz (2006) have argued, teachers' beliefs about teaching and learning with ICT are central to integration, revealing a strong relationship between teachers' educational beliefs, their planning, instructional decisions and classroom practices. It is imperative that these factors need to be included in any professional learning (PL) program that hopes to encourage integration, transformation, and sustainability. "There is much more involved in technology integration beyond acquiring technical skills" (Hixon & Buckenmeyer, 2009, p.140). Gagné claimed (1985) that attitude and affective knowledge requires learning that leads to a change in behaviour, and has emphasised the need for learner-centred PL where teachers have a strong voice in the process and are actively engaged in the learning (Donovan et al., 2007). A learner-centred approach and active involvement provides a personal and emotional connection and provides a better environment for attitudinal change.

In recent years, the field of educational technology has begun to recognise the complex interactions that affect a teacher's ability to use technology effectively in the classroom. There are many models of technology professional learning; one of the most widely recognised is Technological, Pedagogical, Content, Knowledge (TPACK; Mishra & Koehler, 2006). Edyburn (2013) claims that "the most important features of the TPACK are the overlapping areas that provide a context for understanding the impact of technology" (p. 15).

There are multiple benefits from engaging in well-designed professional learning such as the TPACK model. Firstly if teachers are provided with appropriate professional learning, they will become more confident and will be more likely to include the technology effectively in their pedagogy, which will essentially enhance the sustainability of mobile technology. Secondly, involvement in appropriate professional learning will have an impact on the teaching and learning of students with disabilities, enabling classroom teachers to capitalise on the numerous advantages that the literature has demonstrated in the previous section.

Conclusions and recommendations

It is apparent that mobile technology possesses the capability to provide students with disabilities support to gain a variety of skills, knowledge and understandings in academic curriculum, communication, and social and behavioural areas.

Mobile technology can address many of the challenges associated with traditional assistive technologies such as cost, lack of portability, separate devices for each need, and complicated requirements for the devices' use and maintenance. In fact, they have now become *the assistive technology of choice* for students with special needs. The ability to access this support any time and anywhere allows the students with various disabilities to participate more fully in a variety of environments such as an inclusive classroom setting, home and community. This technology is now widely accepted among all the members of modern society and the use of them by students with special needs is supportive of the inclusion paradigm.

There is already some research evidence to support the claims that the use of mobile technology benefits students with special needs in achieving educational outcomes and participating more fully in all school environments. Nevertheless, more research is needed to further explore the impact of mobile technologies for inclusive education and to increase the quality of life of students with disabilities. Potential directions for research include: evaluation of different apps for students with disabilities; ways to utilise mobile technology in improving home-school collaboration; and the quality of professional development for teachers. Larger scale quantitative studies in the area should also be undertaken, in order to provide a more solid evidence base for the use of mobile technology to support students with disabilities.

The use of mobile technology is not a panacea either for students with disabilities or their teachers. However, it shows strong promise in supporting students with disabilities to succeed in inclusive school environments and to develop their potential. The UDL framework combined with mobile technologies are suitable for improving the quality of life for students with special needs.

References

Achmadi, D. (2010). *Teaching a multi-step requesting sequence to two adolescents with autism using an iPod-based speech generating device.* (Unpublished master's thesis). Victoria University of Wellington, Wellington, NZ.

AIHW (Australian Institute of Health and Welfare). (2008). *Disability in Australia: Intellectual disability*. Bulletin no. 67. Cat. no. AUS 110. Canberra: AIHW.

An, Y. J., & Reigeluth, C. (2012). Creating technology-enhanced, learner-centred classroom: K-12 teachers' beliefs, perceptions, barriers and support needs. *International Society for Technology in Education*, *28*(2), 54–62.

Aslan, S., & Reigeluth, C. M. (2011). A trip to the past and future of educational computing: Understanding its evolution. *Contemporary Educational Technology*, *2*(1), 1–17.

Authority 20 U.S.C. (2004). *The individuals with disability education act*.

Ayres, K. M., & Langone, J. (2008). Video supports for teaching students with developmental disabilities and autism: Twenty-five years of research and development. *Journal of Special Education Technology*, *23*(3), 1–8.

Balanskat, A., Blamire, R., & Kefala, S. (2006). *A review of studies of ICT impact on schools in Europe*. European Schoolnet, Brussels, Belguim.

Banda, D. R., Dogoe, M. S., & Matuszny, R. M. (2011). Review of video prompting studies with persons with developmental disabilities. *Education and Training in Autism and Developmental Disabilities*, *46*(4), 514–527.

Bandura (1997). *Self-efficacy: The exercise of control*. Freeman: New York.

Beggs, T. A. (2000). Influences and barriers to the adoption of instructional technology. Paper presented at the *Proceedings of the Mid-South Instructional Technology Conference*, Murfreesboro, Tennessee.

Bingimlas, K. A. (2009). Barriers to the successful integration of ICT in teaching and learning environments: A review of the literature. *Eurasia Journal of Mathematics, Science and Technology Education*, *5*(3), 235–245.

Braddock, D., Rizzolo, M., Thompson, M., & Bell, R. (2004). Emerging technologies and cognitive disability. *Journal of Special Education Technology*, *19*(4), 49–56.

Brown, D. (2011). Some uses of educational and assistive technology for people with disabilities. *Computers & Eductation*, *56*, 1.

Burton, C. E., Anderson, D. H., Prater, M. A., & Dyches, T. T. (2013). Video self-modeling on an iPad to teacher functional math skills to adolescents with autism and intellectual disability. *Focus on Autism and Other Developmental Disabilities*, *28*(2), 67–77.

Campigotto, R., McEwen, R., & Denmans Epp, C. (2013). Especially social: Exploring the use of an iOS application in special needs classrooms. *Computers & Education*, *60*, 74–86.

CAST. (2011a). *About UDL*. Retrieved on December 5, 2014 from http://www.cast.org/udl/index.html

CAST. (2011b). *Universal design for learning guidelines – Version 2.0*. Wakefield, MA: Author. Retrieved on March 10, 2014 from http://www.udlcenter.org/aboutudl/udlguidelines/downloads

Charlop, M. H., Gilmore, L., & Chang, G. T. (2008). Using video modeling to increase variation in the conversation of children with autism. *Journal of Special Education Technology*, *23*(3), 47–66.

Cihak, D., Fahrenkrog, C., Ayres, K., & Smith, C. (2010). Use of video modeling via a video iPod and a system of least prompts to improve transitional behaviors for students with autism spectrum disorders in the general education classroom. *Journal of Positive Behavior Interventions*, *12*(2), 104–115.

Cobb, A. (2010). To differentiate or not to differentiate? Using Internet-based technology in the classroom. *The Quarterly Review of Distance Education*, *11*(1), 37–45.

Courey, S. J., Tappe, P., Siker, J., & LePage, P. (2013). Improved lesson planning with universal design for learning (UDL). *Teacher Education and Special Education: The Journal of the Teacher Education Division of the Council for Exceptional Children*, *36*(1), 7–27.

Cumming, T., Draper Rodriguez, C., & Strnadová, I. (2013). Aligning iPad applications with evidence-based practices in inclusive and special education. In S. Keengwe (Ed.), *Pedagogical applications and social effects of mobile technology integration* (pp. 55–78). Hershey, PA: IGI Global.

Cumming, T., Higgins, K., Pierce, T., Miller, S., Tandy, R., & Boone, R. (2008). Social skills instruction for adolescents with emotional disabilities: A technology-based intervention. *Journal of Special Education Technology*, *23*(1), 19–33.

Cumming, T., & Strnadová, I. (2012). The iPad as a pedagogical tool in special education: Promises and possibilities. *Special Education Perspectives*, *21*(1), 34–46.

Commonwealth of Australia. (2006). *Disability standards for education 2005*. Retrieved on December 3, 2014 from http://www.dpi.vic.gov.au/DPI/nrenfa.nsf/v/96545B5B3372AAE2CA2577AB001EB8FE/file/Disability_Standards_for_Education_2005_Access.pdf

Department for Education and Skills (2001). *Special educational needs code of practice*. London: HMSO.

Dixon, R. M., & Verenikina, I. (2013). Exploring the 'tool metaphor' for using digital technology in teaching students with autism spectrum disorders (ASD). In W. Midgley, K. Trimmer, & A. Davies (Eds.), Metaphors for, in and of Education Research (pp. 156–172). Newcastle upon Tyne: Cambridge Scholars Publishing.

Donovan, L., Hartley, K., & Strudler, N. (2007). Teacher concerns during initial implementation of a one to one laptop initiative at the middle school level. *Journal of Research on Technology in Education*, *39*(3), 263–286.

Douglas, K. H., Wojcik, B. W., & Thompson, J. R. (2011). Is there an app for that? *Journal of Special Education Technology*, *27*(2), 59–70.

Edyburn, D. L. (2010). Would you recognize universal design for learning if you saw it? Ten propositions for new directions for the second decade of UDL. *Learning Disability Quarterly*, *33*(1), 33–41.

Edyburn, D. L. (2013). Critical issues in advancing the special education technology evidence base. *Exceptional Children*, *80*(1), 7–24.

Ertmer, P. A., Ottenbreit-Leftwich, A. T., Sadik, O., Sendurer, E., & Sendurer, P. (2012). Teacher beliefs and technology integrations practices: A critical relationship. *Computers & Education*, *59*, 423–435.

European Commission. (2013). *Support for children with special educational Needs*. Brussels: European Commission.

Farrell, J. (2013). iPhone/iPad apps for AAC. *Spectronics inclusive learning technologies*. Retrieved on May 24, 2014 from https://www.spectronicsinoz.com/iphoneipad-apps-foraac

Fernández-López, A., Ródriguez-Fortíz, M., & Ródriguez-Almendros, M. (2013). Mobile learning technology based on iOS devices to support students with special education needs. *Computers & Education*, *61*, 77–90.

Finson, K. D., & Ormsbee, C. K. (1998). Rubrics and their use in inclusive science. *Intervention in School and Clinic*, *34*(2), 79–88.

Flores, M., Musgrove, K., Renner, S., Hinton, V., Strozier, S., Franklin, S., & Hil, D. (2012). A comparison of communication using the Apple iPad and a picture-based system. *Augmentative and Alternative Communication*. Advance online publication. doi: 10.3109/07434618.2011.644579

Ford, C., & Rabe, K. (2011). Using the iPhone for assistive technology: A case study. *Exceptional Parent*, 20–22.

Gagné, R.M. (1985). *The conditions of learning and theory of instruction*. New York: Holt, Rinehart, and Winston.

Gargiulo, R. & Metcalf, D. (2010). *Teaching in today's inclusive classrooms: A universal design for learning approach*. Belmont, CA: Wadsworth.

Gentry, T., Lau, S., Molinelli, A., Fallen, A., & Kriner, R. (2012). The Apple iPod touch as a vocational support aid for adults with autism: Three case studies. *Journal of Vocational Rehabilitation*, *37*(2). Advance online publication doi: 10.3233/JVR-2012-0601

Gentry, T., Wallace, J., Kvarfordt, C., & Lynch, K. (2010). Personal digital assistants as cognitive aids for high school students with autism: Results of a community-based trial. *Journal of Vocational Rehabilitation*, *32*(2), 101–107.

Hall, T. (2009). *Explicit instruction*. Wakefield, MA: National Center on Accessing the General Curriculum. Retrieved March 10, 2014 from http://aim.cast.org/learn/historyarchive/backgroundpapers/explicit_instruction.

Hall, T., Strangman, N., & Meyer, A. (2011). *Differentiated instruction and implications for UDL implementation*. Wakefield, MA: National Center on Accessing the General Curriculum. Retrieved March 28, 2012 from http://aim.cast.org/learn/historyarchive/backgroundpapers/differentiated_instruction_udl

Hammond, D., Whatley, A., Ayres, K., & Gast, D. (2010). Effective use of video modelling to teach iPod use to students with moderate intellectual disabilities. *Education and Training in Autism and Developmental Disabilities*, *45*(4), 525–538.

Hixon, E., & Buckenmeyer, J. (2009). Revisiting technology integration in schools: Implications for professional development. *Computers in Schools*, *26*(20), 130–146.

Hopkins, I.M., Gower, M.W., Perez, T.A., Smith, D.S., Amthor, F.R., Wimsatt, F.C., & Biasini, F.J. (2011). Avatar assistant: Improving social skills in students with an ASD through computer-based intervention. *Journal of Autism and Developmental Disorders*, *41*(11), 1543–1555.

Hourcade, J. P., Bullock-Rest, N. E. & Hansen, T. E. (2012). Multitouch tablet applications and activities to enhance the social skills of children with autism spectrum disorders. *Personal and Ubiquitous Computing*, *16*(2), 157–168.

Jeffs, T., Behrmann, M., & Bannan-Ritland, B. (2006). Assistive technology and literacy learning: Reflections of parents and children. *Journal of Special Education Technology*, *21*(1), 37–44.

Johnson, L., Adams, S., & Haywood, K. (2011). *The NMC Horizon Report: 2011 K-12 edition*. Austin, TX: The New Media Consortium.

Jowett, E. L., Moore, D. W., & Anderson, A. (2012). Using iPad based video modelling package to teach numeracy skills to a child with autism spectrum disorder. *Developmental Neurorehabilitation*. doi:10.3109/17518423.2012.682168.

Kagohara, D.M. (2011). Three students with developmental disabilities learning to operate an iPod to access age-appropriate entertainment videos. *Journal of Behavioral Education*, *20*, 33–43.

Kagohara, D. M., van der Meer, L., Achmadi, D., Green, V. A., O'Reilly, M. F., Lancioni, G. E., et al. (2012). Teaching picture naming to two adolescents with autism spectrum disorders using systematic instruction and speech-generating devices. *Research in Autism Spectrum Disorders*. doi:10.1016/j.rasd.2012.04.001.

Kagohara, D.M., van der Meer, L., Achmadi, D., Green, V. A., O'Reilly, M. F., Mulloy, A., et al., (2010). Behavioural intervention promotes successful use of an iPod-based communication device by an adolescent with autism. *Clinical Case Studies*, *9*, 328–338.

Kagohara, D. M., van der Meer, L., Ramgoss, S., O'Reilly, M. F., Lanioni, G. E., Davis, T. N., Rispoli, M., Lang, R., Marschik, P. B., Sutherland, D., Green, V. A., & Sigafoos, J. (2013). Using iPods and iPads in teaching programs for individuals with developmental disabilities: A systematic review. *Research in Developmental Disabilities*, *34*(1), 147–156.

Koetsier, J. (2013). 700K of the 1.2M apps available for iPhone, Android and Windows are zombies. *VB News*. Retrieved on December 7, 2014 from http://venturebeat.com/2013/08/26/700k-of-the-12m-apps-available-for-iphone-android-and-windows-are-zombies/

McGuire, J., Scott, S., & Shaw, S. (2006). Universal Design and its application in educational environments. *Remedial and Special Education*, *27*, 166–175.

Mechling, L. C., Gast, D. L., & Krupa, K. (2007). Impact of SMART board technology: An investigation of sight word reading and observational learning. *Journal of Developmental Disorders*, *37*(10), 1869–1882.

Merbler, J., Hadadian, A., & Ulman, J. (1999). Using assistive technology in the inclusive classroom. *Preventing School Failure*, *43*(3), 113–118.

Mishra, P., & Koehler, M.J. (2006). Technological pedagogical content knowledge: A framework for integrating technology in teachers' knowledge. *Teachers College Record*, *108*(6), 1017–1054.

Mumtaz, S. (2006). Factors affecting teachers' use of information and communications technology, a review of the literature. *Journal of Information Technology for Teacher Education*, *9*(3), 319–342.

Neely, L., Rispoli, M., Carmago, S., Davis, H., & Boles, M. (2013). The effect of instructional use of an iPad on challenging behaviour and academic engagement for two students with autism. *Research in Autism Spectrum Disorders*, *7*(4), 509–516.

Newton, D. A., & Dell, A. G. (2011). Mobile devices and students with disabilities: What do best practices tell us? *Journal of Special Education Technology*, *26*, 47–49.

Nunley, K. (2006). *Differentiating the high school classroom: Solution strategies for 18 common obstacles*. Thousand Oaks, CA: Corwin Press.

Nunley, K. (2003). Layered curriculum brings teachers to tiers. *Education Digest*, *69*(1), 31–36.

Ozdemir, S., Univeritesi, G., Fakultesi, G. E., & Bolumu, O. E. 2008. Using multimedia social stories to increase appropriate social engagement in young children with autism. *Turkish Online Journal of Educational Technology*, *7*(3), 80–88.

Palak, D., Walls, R., & Wells, J. (2006). Integration of instructional technology: A device to deliver instruction or a tool to foster learning. *International Journal of Instructional Media*, *33*(4), 355–365.

Pelgrum, W. J., (2001). Obstacles to the integration of ICT in education: results from a worldwide educational assessment. *Computers in Education*, *37*(2), 163–178.

Price, A. (2011). Making a difference with smart tablets: Are iPads really beneficial for students with autism? *Teacher Librarian*, *39*(1), 31–34.

Sailers, E. (2010). How schools can successfully implement iPod touch and iPad for children with special needs, *Closing the Gap*, *29*(4), 8–10.

Schoepp, K. (2005). Barriers to technology integration in a technology rich environment. *Learning and Teaching in Higher Education: Gulf Perspectives*, 2(1), 1–24.

Seok, S., DaCosta, B., Kinsell, C., Poggio, J., & Meyen, E. (2010). Computer-mediated intersensory learning model for students with learning disability. *TechTrends*, *54*(2), 63–71.

Shane, H.C., Laubscher, E.H., Schlosser, R.W., Flynn, S., Sorce, J.F. & Abramson, J. (2012). Applying technology to visually support language and communication in individuals with Autism Spectrum Disorders. *Journal of Autism and Developmental Disorders*, 42(6), 128–1235.

Shrestha, A., Anderson, A. & Moore, D.W. (2013). Using point-of-view video modelling and forward chaining to teach a functional self-help skill to a child with autism. *Journal of Behavioural Education*, 22(2), 157–167.

Stafford, L. (2011). Technology transforms the autism classroom of the future. *Autism Spectrum Quarterly*, 4, 7–10.

Stanford, P., Crowe, M., & Flice, H. (2010). Differentiating with technology. *Teaching Exceptional Children Plus*, 6(4), 2–9.

Tomlinson, C. A. (2008). The goals of differentiation. *Educational Leadership*, 66(3), 26–30.

van der Meer, L., Kagohara, D.M., Achmadi, D., O'Reilly, M.F., Lancioni, G.E. et al. (2011). Teaching functional use of an iPod-based speech-generating device to individuals with developmental disabilities. *Journal of Special Education Technology*, 26(3), 1–11.

van der Meer, Diden, R., Sutherland, D., O'Reilly, M.F., Lancioni, G.E., & Sigafoss, J. (2012a). Comparing three augmentative and alternative communication modes for children with developmental disabilities. *Journal of Developmental and Physical Disabilities*. http://dx.doi.org/10.1007/s10882-012-9283-3

van der Meer, L., Sutherland, D., O'Reilly, M. F., Lancioni, G. E., & Sigafoos, J. (2012b). A further comparison of manual signing, picture exchange, and speech-generating devices as communication modes for children with autism spectrum disorders. *Research in Autism Spectrum Disorders*. doi:10.1016/j.rasd.2012.04.005.

Vedora, J., & Stromer, R. (2007). Computer-based spelling instruction for students with developmental disabilities. *Research in Developmental Disabilities*, *28*(5), 489–505.

Verenikina, I., Tanner, K., Dixon, R. M. & de Graaf, E. (2010). Interactive whiteboards as a tool for teaching students with autism spectrum disorders. In Howard, S. Katherine. (Eds.). *AARE International Education Research Conference*. Melbourne: AARE.

Walker, H. (2011). Evaluating the effectiveness of apps for mobile devices. *Journal of Special Education Technology*, 26(4), 59–63.

Wehmeyer, M., Palmer, S., Smith, S., Davies, D., & Stock, S. (2008). The efficacy of technology use by people with intellectual disability: A single-subject design meta analysis. *Journal of Special Education Technology*, 23(3), 21–30.

Wilson, M. (2011). APPSolutely accommodating. *Journal of Special Education Technology*, 26(2), 55–60.

Xin, J., & Sutman, F. (2011). Using the smart board in teaching social stories to students with autism. *Teaching Exceptional Children*, *43*(3), 8–24.

8 Sustaining learning with mobile devices through educational design for teaching presence

Caroline Walta and Howard Nicholas

Introduction

Sharples, Taylor and Vavoula (2007, p. 1) define mobile learning as "the processes of coming to know through conversations across multiple contexts amongst people and personal interactive technologies". Learning in such contexts is created through interaction, which requires support through portable and ubiquitous technologies. In teachers' discussions of issues surrounding sustainability in mobile learning, sustainability is often associated with pedagogical beliefs related to the learning context (Sharples et al., 2007). Other issues identified include ongoing access to funding, the presence or absence of professional ICT support and the need for leaders who encourage and support learning initiatives that build on ICT and innovation in curriculum (Cochrane, 2010; Herrington et al., 2008; Ng & Nicholas, 2013).

It often appears to be the case that these issues are identified in relation to one particular device, particularly if there is only one kind of mobile device carried by the learners. However, the mobile device is just one component of a suite of technologies. These technologies include wireless routers, servers, printers, desktop (or in some circumstances laptop) computers and, increasingly as 'Bring Your Own Device' (BYOD) initiatives become more widespread, different kinds of mobile devices with varied affordances. An additional element is the diversity in the student group, particularly in terms of prior experiences with and formed attitudes toward mobile devices. In the context of the kind of program that we discuss in this chapter, a further issue is variation in resources available to the learners in places other than their own formal educational setting. Sustaining programs where mobile devices are used for learning therefore involves careful design to embrace the extensive and dynamic diversity of learners and their technologies in ways that bring the diverse elements into a shared space and provide a sense of common purpose.

Claims about the significance of the above elements for the sustainability of innovative initiatives are equally present in schools and in higher education. We will discuss the vital importance of learning design to enable teaching and learning in a mobile technological environment to remain the key outcome of participation in the program. We analyse our experiences of six years of involvement in a graduate-level teacher education program that continues to be delivered in blended learning mode and features the inter-dependent use of multiple technological devices as an integral component of the learning environment and process. Mobile technology was central to the development of this program. We show that principled but flexible design was a key to sustaining the use of mobile technology.

The term *mobile technology* for the purposes of this chapter encompasses the suite of devices that enables potentially simultaneous access to multiple (including remote) sources of information as well as multiple means of communication between the learners themselves as well as with other participants (see the critique of 'container-like' metaphors in Leander, Phillips, & Headrick Taylor, 2010). Our framing is consistent with Sharples et al. (2007) in that it emphasises the ongoing interaction between users and technologies but our framing highlights the notion of a suite of technologies and the exploitation of multiple sources of information that may vary with context. In contrast, Sharples et al. (2007) focus more on mobile devices without attention to fixed resources such as desktop computers and printers. A corollary of our framing is that learners need to be able to be both independent and interdependent. They also need to be able to exercise critical capacity. Learners need to be able to access information (technical control) and evaluate that information (cognitive and emotional control) (see Ng, 2012, p. 56 for a view of the dimensions of digital literacy). This sets challenges if learners come with different (and sometimes negative) experiences of or attitudes toward the use of mobile technology in general and mobile devices (for learning) in particular.

In a climate of socially-present, media-enabled communication and widespread use of a variety of technological devices, the challenge for teacher educators is to consider the capacity for technologies to optimise learning in light of the substantial variation in devices, learning experiences, purposes for engagement and contexts of engagement. Education, in these reshaped contexts (since we can no longer presume to speak of only one context) should be seen as learning through thinking in association with multiple sources of information, values, experiences and opportunities for (different kinds of) interaction. The challenge for teacher educators is to design curriculum that reflects current beliefs about creating challenging thinking environments for learning. This learning needs to occur in a socially constructivist environment which offers the capacity for learning anywhere and any time. Students (i.e. both these pre-service teachers themselves and those students who will be taught by the pre-service teachers in the program) expect to be able to engage with multiple sources of information while being connected both socially and academically to fellow students and academic mentors (Laurillard, 2007). Pachler, Bachmair and Cook (2010, p. 5) refer to the work of Kress and Pachler (2007, p. 22) when they propose the following outcome in line with this expectation:

> One defining feature of mobile learning for us is the need for individuals to go beyond the acquisition of knowledge relevant to issues encountered in the world but also to shape their knowledge out of their own sense of their world.

It is unfair to leave finding ways to address this complex and changing set of expectations to the moment-by-moment decisions of teachers since the necessary resources will not be present to support such decisions. Instead, effective design processes need to have prepared for the presence of or access to these resources (see the discussion of designing for the role of teachers in Goodyear and Dimitriadis, 2013).

Sustainability in this context is in the hands of curriculum designers. Their work needs to be informed by theories of learning, knowledge of effective program design, knowledge of the content of the curriculum and generic knowledge of mobile technological capabilities (see discussion in Kearney et al., 2012). This establishes a sound theoretical basis within which improved learning outcomes can be supported through technological change and innovation. Goodyear and Dimitriadis (2013, no page) note that:

> Crucially, the design work that dominates the life of a project tends to underprepare for: (1) the complexities of embedding or enacting the design in new contexts, (2) the customisation or co-configuration that is done by learners and teachers to make the innovation work in new contexts, (3) supporting the work of teachers who have to make the innovation function, after the end of the funded project, and (4) periodic review, modification and/or redesign of the innovation.

The curriculum designers need to consider not only the content of what is being taught, but the widest sense of the processes that are being used to engage with that content, including therefore the technology as a shaper of both the processes and the content. Sharples, Taylor and Vavoula (2007, p. 3) draw on the work of the United States National Research Council in describing effective learning environments as centred around: the learner; sound and creative perspectives on knowledge; assessment matched to the needs of the learners and a community that is mutually supportive. Korthagen, Loughran and Russell (2006) support these contentions in putting forward a number of principles for effective teacher education, which include a call for the interrogation of conflicting demands, a view of knowledge as a subject to be created and a focus on the learner rather than the curriculum. Building on these ideas, we conclude that effective course design should ensure that the content that is explored is:

- relevant (for example ensuring that images of classrooms and demonstrations of learning/teaching processes encompass current appearances, resources and leading-edge practices);

- layered (offering opportunities for multiple and sometimes conflicting perspectives to be explored).

The processes that are demonstrated/experienced include:

- a variety of practices covering the range of modalities of interaction (Nicholas & Starks 2014);
- content of different kinds introduced through both mastery-oriented activities and exploratory activities and providing ways of engaging with the reasoning behind such designs.

The technology that is presented/used:

- does not overwhelm learners with breadth only for the sake of breadth;
- is not only used, but also is supported by opportunities for learning how to use it effectively;
- reduces pressure on individuals to constantly master technological innovation in isolation by supporting processes where all participants are partners in the learning process;
- provides ways in which individual learners can move into the role of 'teacher' either through expertise in one of the presented applications or by sharing expertise in an application/device that was not presented.

The reflections on some of the experiences of staff and students in a single pre-service teacher education program as they negotiated working across the diverse technologies over multiple iterations of the same one-year long program come from a study approved by the La Trobe University Ethics Committee. All data has been made anonymous.

A description of the program

The Graduate Diploma in Education Middle Years program is a one-year graduate-entry program at La Trobe University in Victoria, Australia. Students who have completed at least their first university degree can use this program to qualify to teach across all sectors of school education in Victoria (P–12). 'Middle years' refers to the period between Year 5 and Year 9/10, when students are aged between 11 and 16. Students entering this program are informed that four weeks will be spent in activities in intensive face-to-face mode at the Shepparton Campus, some 200 km from the main campus in Melbourne. The remaining weeks of their academic studies (excluding the nine weeks of practicum) are managed in an online environment. In the description below, a key purpose is to outline how the technologies and the practices have evolved over time as (a) lecturers have gathered experience (and feedback) about what works and (b) technologies have changed, both in terms of changes within particular applications and in terms of the university's approach to technology as part of standard teaching and learning. Some of this has been a process of untheorised adjustment as we have sought to make the best of changes that were beyond our control. One constant principle has been that our students should learn about the use of technologies through the (supported and reflective) experience of using them and seeing them used for teaching. A second principle has been that the uses and technologies should be open to critique. A third principle has been that uses should be collaborative or in the context of student–student collaboration. Consistent focus has been given to the use of technologies in assessment tasks that form part of the course.

Beginning in 2008, the program has been taught using a combination of face-to-face intensive workshops on campus and various e-learning technologies. The main technologies are the university's learning management system (then, *Blackboard* and now *Moodle*) and the virtual classroom software (initially *Elluminate* and now *Blackboard Collaborate*). Increasing use has been made of *PebblePad*, an e-portfolio application that has been progressively integrated with the learning management system, but also contains its own assessment submission and feedback system. This combination of technologies enables both synchronous and asynchronous interaction while students are not on the campus. Initially, the program did not involve mobile technologies although both the students and the staff were highly mobile in the course of the year.

The iPod initiative commenced in 2010. In the second half of 2009, special funding had enabled the purchase of sufficient wireless enabled iPod Touches that each enrolled student could be lent one for the duration of the program (approximately nine months). The funding also permitted the purchase of a server to support greater integration of mobile learning to complement the affordances of the other technologies. Other students had access to the same devices in ensuing years but progressively the pool of devices has broadened as students have predominantly brought their own and the iPods became used more as class-sets that could be used with/by school students during teaching practicum placements.

The Mac OS X Server (henceforth, the server) primarily supported podcasting and as the program moved from a pilot (2010) to a standard part of the university's teaching approach, more general university-wide technologies came to replace it, initially *Lectopia* then *Echo 360* and subsequently *Echo System*. These changes complemented changes in the university's philosophy about the use of learning management systems as greater acceptance emerged of two-way interaction between lecturers and students in the negotiation of course content.

The attachment of the server to the university system in 2010 enabled staff involved with the program to use *Podcast Capture* to generate and place podcast material onto the server. Students were then able to download this material to their iPods through an RSS feed or work directly from the server on their computers. The students were also able to create, post and access material to share via the server.

The podcasting associated with the server was an alternative, flexible platform for generating, sourcing and accessing audio and multimedia material relevant to learning about teaching and provided the pre-service teachers with opportunities to consider the potential of hand-held technologies in their own learning and also in their teaching. The server was used both as a means to distribute pre-prepared activities and to provide access to the steady diet of community/media comment on educational issues. Used in this way, the server was a means of accessing and contributing to student exposure to current information and opinion exchange. Material was sourced from print and digital media comment on and research into educational issues. In the initial year of the iPod initiative, reflection included community and political debate around issues such as the introduction of the national curriculum or school rating websites that the then national government was introducing. The discussions occurred during the weekly virtual classroom-based tutorials. Increasingly over the years the use of podcasts as part of virtual classroom work has replaced more lecturer-fronted presentations. While these podcasts are still 'knowledge-delivery', the delivery is now in much smaller chunks surrounded by opportunities for discussion and alternative input.

Principles underpinning the program design

In the past decade there has been a proliferation of studies that have sought to establish the essential elements of effective online learning environments for students (Shackelford & Maxwell, 2012). As universities turn ever more frequently to online and blended modalities for course delivery with the affordances of new, increasingly ubiquitous and powerful communication technologies, a key focus of research has become the ability of these modalities to create and sustain communities of inquiry (Garrison, Anderson, & Archer, 2000; Garrison, 2011; Hardman, 2008). The concept of a Community of Inquiry (Garrison, Anderson, & Archer, 2000) is a useful framework for exploring sustainability in learning with mobile devices since it contains dynamic elements that are needed to accommodate the shifting relationships between learners and teachers, particularly in programs that recur on an annual basis with the 'same' teachers, but new students and shifting technologies.

Both Hiltz and Wellman (1997) and Dettori, Giannetti and Persico (2006) argued for a definition of a community framed by joint purpose in a virtual environment rather than one framed by temporal space. This is consistent with recent research showing that well-designed courses can support communities of connected students in virtual environments (Lemon, 2013, 2015; Rovai, 2002).

Figure 8.1 depicts Garrison's (2011, p. 23) description of the Community of Inquiry (CoI) model.

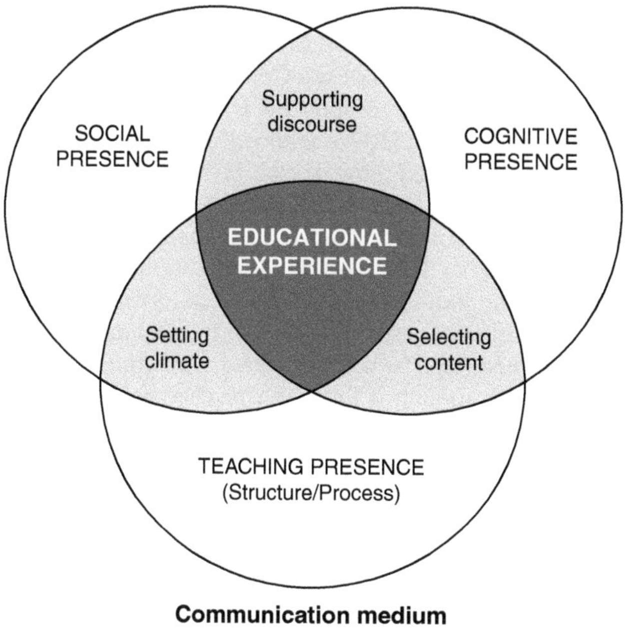

Figure 8.1 Community of Inquiry.

In the CoI model (Garrison, 2011, p. 30), social presence reflects and promotes forms of interaction that create a sense of belonging and support collaborative learning. Garrison (2011, p. 127) argues that social presence contributes to learning and interaction through its three elements of interpersonal communication, open communication and cohesive communication (Garrison, 2011, p. 38–9). Cognitive presence is "the intellectual environment that supports sustained critical discourse and higher order knowledge acquisition and application" (p. 42). Teaching presence involves

> shaping the appropriate transactional balance and, along with the learners, managing and monitoring the achievement of worthwhile learning outcomes in a timely manner. … [It] performs an essential service in identifying relevant societal knowledge, designing experiences that will facilitate reflection and discourse and diagnosing and assessing learning outcomes.
>
> (Garrison, 2011, p. 54–5)

Cognitive presence is the component reflecting the ability of a community of learners to construct meaning through sustained communication. It reflects the students' developing understanding as they interact and collaborate (Shackelford & Maxwell, 2012, p. 2).

The Community of Inquiry model has generated substantial interest among online researchers as indicated in a review of issues and future directions by Garrison and Arbaugh (2007). Garrison and Arbaugh (2007) noted that research had focused on the three elements of the CoI model separately even though the interrelationship between the components is the model's essential concept (p. 159). Garrison and Arbaugh (2007) acknowledged the research of Brown (2001), which had highlighted how social presence evolves and becomes more complex (p. 160). The authors concluded that the purpose of social presence is to create conditions for inquiry and interaction through reflective and threaded discussions so that "worthwhile educational purposes" can be achieved (p. 160). In the model, cognitive presence, defined as "the extent to which learners are able to construct and confirm meaning through sustained reflection and discourse" (p. 160) is associated with course design. Course design is the vehicle for forging a link with teacher presence since course design is one component of teachers' work. Research suggests a complementary relationship between teaching presence and cognitive presence.

> Social presence lays the groundwork for higher level discourse and the structure, organisation and leadership associated with teaching presence creates the environment where cognitive presence can be developed.
>
> (Garrison & Arbaugh, 2007, p. 163)

Teacher presence is linked to social presence when social presence involves learning through a community, facilitated through conscious course design (Pallof & Pratt, 2003). This finding was supported by Pawan et al. (2003), who found that teaching presence was reflected in the depth and quality of both students' interactions and their discussions. Other research has linked teacher presence to cognitive presence by making the link between course design and the capacity for students to construct meaning through social interaction (Swan et al., 2008).

Shea, Li and Pickett (2006) drew on the work of Chickering and Gamson (1987) to affirm that good learning environments "promote high levels of interaction between students and faculty, time on task, prompt feedback, active roles for learners, a greater emphasis on student cooperation over competitiveness and a respect for diversity in approaches to learning" (Shea et al., 2006, p. 176). They concluded that teacher presence influences students' sense of a learning community. Shea et al. (2006, p. 177) and Garrison (2011) argued for the importance of effective feedback and course design in fostering all three components of a CoI. This finding reflects the original arguments of Garrison, Anderson and Archer (2000, p. 88) that "… Community of Inquiry assumes that learning occurs within the Community through the interaction of three core elements …: cognitive presence, social presence, and teaching presence".

Overall there is widespread agreement in the works cited that it is critical to embed teacher presence in the design of an effective online learning environment. Among other reasons, this is because students attach value to clearly defined course requirements and timely feedback (Brinkerhoff & Koroghlanian, 2007; Durrington et al., 2006; Sheridan & Kelly, 2010, p. 7). This chapter looks at these factors as elements contributing to the sustainability of the mobile learning innovation over six years. The data were obtained from surveys conducted at the end of the first two years of the program, individual and focus group interviews with students and staff, conducted at the end of the second year of the program and through analysis of open comments in the discussion forums that were part of the program.

Theoretical issues in program design

The need for program design to align outcomes with learning activities supported by technology is emphasised in current literature (Beetham & Sharpe, 2007; Biggs & Tang, 2007; Oliver, 1999; Salmon & Edirisingha, 2008; Stephenson, 2001). Other important considerations include implementing learning management systems so that they facilitate communication and interaction between students in programs and courses (Garrison & Kanuka, 2004; King, 2002; Rovai, 2002). The set of Web 2.0 applications (podcasts, weblogs, wikis etc.) embedded in the design of the program increases support for learners as active participants, creators of knowledge, and seekers of engaging personal experiences (McLoughlin & Lee, 2008, p. 10) but it is also designed to include teachers and teaching. This reflects a view that rather than focusing on the interactivity of the Web 2.0 tools themselves, it is important to design an environment where learning that is shared by all participants is the focus of the design (Dale & Pymm, 2009; Edirisingha, Salmon, & Fothergill, 2006; Lee & Chan, 2007a, 2007b; McLoughlin & Lee, 2008).

The design for the program discussed here is informed by principles that emphasise the potential for technology to support learning within what has already been identified as a constructivist paradigm. This approach places its emphasis on exploring new options in program design underpinned by established theories about learning through constructing meaning, but sets a crucial role for teaching in initiating and shaping the Community of Inquiry, as outlined by Mayes and Frietas (2007, p.14) who pointed out that "we are witness to a new model of education, rather than a new model of learning".

Technology in enabling a community of inquiry

Table 8.1 links the use of elements of a Community of Inquiry to technological tools which support elements of Social Presence and Cognitive Presence, both of which are linked to connectedness and learning. These elements become the means by which Teacher Presence promotes discourse and critical enquiry. In the third column, the presence of multiple technologies in association with each of the elements shows how the sustainability of the program required change within its suite of technologies. In the final

Table 8.1 Relations between elements of Community of Inquiry and technologies and sustainability

Elements in the Community of Inquiry	*Online activity*	*Technological Tool/ Online learning system*	*Sustainability inherent in model*
Social Presence Interpersonal Open communication	LMS Discussion spaces: Server – personal wikis Practicum blogs – associated with seeking help and sharing experiences *Elluminate* – associated with small group use *Skype* – associated with *iPod* use *Elluminate* for tutorial small group discussions	Learning Management System – *Blackboard* *Elluminate* *Collaborate* *Skype* *iPod Touch; BYOD*	Content material continuously being interrogated and created through shared interaction between all members of the community
Cognitive Presence Facilitating reflection and discourse Higher order thinking	Assessment tasks Reflective and critical inquiry associated with selected learning material – use of collaborative spaces in LMS and server *Elluminate* tutorials	LMS Server *Elluminate* *Collaborate*	Flexible, approaches to accessing and processing of learning material accepted as part of the course design
Teacher presence Shaping the learning experience to promote reflection and discourse Collaboration with learners to achieve learning outcomes	Podcasting as part of design to promote shared learning and support discourse; podcasting in response to assessment tasks; used to generate audio files of interviews for posting and tutorial discussion; students source and share online podcasts and other resources	*iPod Touch* associated with server Evolving use of personal devices especially smart phones and iPads	Use of collaborative spaces to promote shared learning

column of the table, comments show how the design of the program contributed to its sustainability for each element of the model.

Activities in the program reflect a focus on collaboration, authentic learning experiences and higher order thinking, especially involving reflection on learning, including:

- sharing observations of classroom approaches to pedagogies and management through open discussions in the LMS and via the server;
- building a portfolio of evidence of learning, including links to resources and software using designated software using *PebblePad*;
- creating units of work through collaboration and sharing online using the LMS;
- creating multimedia podcasts which simulate student learning activities associated with learning about literacy using the iPod Touch;
- creating multimedia podcasts which are themselves learning objects accompanied by audio podcasts about their use in teaching using the iPod Touch;
- participation in supportive blogging during the practicum using both the LMS and the server;
- evaluation of software for educational use in a shared environment;
- opportunities to present and discuss components of a developing philosophy of education with lecturers and other colleagues using an interactive online module featuring shared discussion/journal spaces;
- responding reflectively to online readings and other sourced material in the LMS;
- engaging in peer-to-peer review of teaching where physical location allows for this;
- supporting others though tutorial blogs during practicum using both the LMS and the server.

As we have argued elsewhere (Walta & Nicholas 2013), these uses of new technologies do not create a new 'paradigm' of learning. Echoing our earlier work, we show below that the practices adopted in the program are aligned with the seven fundamental principles for teacher education programs outlined by Korthagen, Loughran and Russell (2006). By underpinning course design with fundamental principles for effective teacher education, and therefore providing an explicit place for the teacher and teaching, the sustainability of the program is enhanced. New technologies offer alternative pathways to achieving effective learning that is relevant to and supported by innovation in technology. This practice allows for evolving adaption of a range of technological devices as the function and purpose of the device is in focus rather than the device itself.

In Table 8.2, teacher education elements (Korthagen, Loughran, & Russell, 2006, pp.1025–1036) are matched with the technology in the program that supports them.

Table 8.2 Relations between elements of teacher education and technologies

Learning context	*Technology*	*Sustainability as reflected in evolving uses of technologies*
Shared participation and interaction	*Elluminate* for tutorials; wiki space for podcasting; student blogs; public posting of assessment	Adaptation to a new learning management system, *Moodle*, use of *Blackboard Collaborate* for real time connection and sharing; use of personal mobile devices, *YouTube*, *Vimeo* for sharing multimedia
Opportunities for reflection	Reflective journals in the LMS; tasks associated with practicum e.g. reflecting on audio recording of teaching; reflecting on feedback from student surveys on teaching	
Opportunities for collaboration	Joint participation in assessment utilising online technologies for collaboration; use of *Elluminate*	Working with peers on collaborate assessment tasks using *Blackboard Collaborate*
Opportunities to receive/respond to feedback/scaffolding learning	Structure of assessment tasks in the LMS; use of designated blogs; public sharing of assessment in the LMS and on *Wiki spaces*	
Opportunities to learn about and through technologies	Some autonomy associated with assessment; public sharing; community of learners enabled to assist in an online environment	
Engagement with authentic curriculum through structure of the learning modules, podcasting and the nature of assessment items	All assessment tasks were learning activities designed for learning about teaching and modelling approaches for future classroom use – e.g. creating podcasts; digital stories associated with critical literacy development in schools, evaluating the potential for use of a range of technologies in student learning	Potential to alter the nature of technological expectation associated with assignment presentation as the new software became available
Opportunities to be knowledge creators	Use of podcasting – both created and sourced placed in wiki space for discussion or written comment	Expanded to use of *YouTube* for presenting material both created and sourced
Opportunities to form communities of learners in a virtual environment	Use of virtual tutorial groups associated with podcasting, meeting regularly in *Elluminate* (later *Collaborate*) with associated reciprocal peer review in practicum placement	

Data obtained from course evaluation surveys and focus groups demonstrated that students were responsive to the use of short podcasts as introductory and stimulus material to be followed up in online collaborate discussions. Students were directed to listen to certain material, often containing conflicting opinions and approaches and to consider certain questions for discussion in their online tutorial that week. Other early initiatives that reflected this approach and therefore were extended into later years using alternative technologies included directing students to gather data themselves and post it on the server for others to consider. This data often took the form of recorded interviews and links to podcasts from *YouTube*, *EdPod* and other sources accessed by students at their own initiative. This process, which aligns with the research of Korthagen et al. (2006) is sustained through evolving technologies associated with shared sites for storage of multimedia and use of personal mobile devices for data production. Use of cloud technology and camera-recording facilities greatly enhanced student capacity to access, create and store material for sharing, thus improving the capacity for students to be knowledge creators within the program.

During the very first face-to-face intensive component, students attended workshops run by two teachers from schools that featured iPod use in personalised learning programs. These workshops enabled students to become more familiar with educational applications (apps) for the iPods and with their potential within educational programs in schools. A blog was set up in LMS *Blackboard* in which students could share ideas for other educational apps. In later years, we brought back into the program graduates from earlier years to share their stories and to create support networks for new students. The ever evolving applications available for Mac devices present ongoing opportunities for teaching and learning in both higher education and schools, but the continual expansion of options can lead to pressure to chase the latest technology. The concept of a community of learners with shared responsibility for supporting others where appropriate underpins an important element of sustainability of a program with a high dependence on technology for learning. One reason for the pressure is that each teacher feels an individual responsibility to be current. Creating a community of learners (or of teachers) who can critically evaluate and, if appropriate, be experts in specific technologies can reduce some of that pressure. While structures exist within both the face-to-face element of the program and on the LMS to solve problems which may arise, student use of 'private' *Facebook* spaces is increasingly common as a place where they collaborate and seek help away from the jurisdiction of course managers. This process sets up relationships among learners as part of a learning community. As part of a process of bonding, the establishment of such communities can extend into the entry into professional practice after an initial teaching qualification has been obtained. Other technological advances have been the capacity of *Blackboard Collaborate* to be a medium through which students can assist each other at any time of their choosing. In this case, a capacity within the application was made a more regular part of the program once the students had found it useful. Sustainability is ensured through a range of online options for accessing assistance to master new technologies and through a principle of flexibility so that program staff are able to respond to student requests.

A feature of the program up to the current iteration is that collaborative tasks are set where students make decisions about the use of technology to produce and disseminate their product. Some tasks are presented in face-to-face environments but there is always an imperative to share output and learning in a collaborative way for the development of shared expertise. Students make use of designated spaces on the LMS, and through links to their own password-protected areas in sites such as *Vimeo* and *Facebook*.

As students and staff became more familiar with the design options associated with podcasting and the capacity of the server to allow the sharing of learning objects, further uses were trialed. In the first year of the program, a number of semester 2 assessment items involving use of technology were changed to reflect the capacity of the server to host multimedia files from both staff and students. These included students creating an educational app, which targeted an aspect of numeracy or literacy and producing a digital story suitable for middle years literacy teaching. Staff and students also set up communication blogs for students during teaching practicum. The space enabled students to post audio files of their own teaching, images of student work and other material with reflective comments. Sustainability was ensured as the basic principles of a community of enquiry (Garrison, 2011) were applied using ever evolving and more powerful new technologies in learning about teaching.

Podcasting: part of the journey in sustaining the program

Research before 2006 tended to focus on the issue of podcasting as a substitute for face-to-face lecturing and whether this is appropriate in terms of student rights and student learning (Heilesen, 2010, p. 2). Later research, which is informed by the evolution of podcasting from audio to multimedia production with the advent of more sophisticated handheld devices, focuses more on instances where podcasting has become part of the overall design of a learning program, utilising its potential to offer a range of learning opportunities in association with the learning outcomes and other technological tools in the program (Draper & Hitchcock, 2006; Edirisingha, Salmon, & Fothergill, 2006; Evans, 2008; Lee & Chan, 2007a, 2007b; Lonn & Teasley, 2009; McLoughlin & Lee, 2008; Ragusa, Chan & Crampton, 2009).

In line with the philosophical underpinning of the program, where some content material was constructed through dialogue and access to competing sources of knowledge, use of podcasting in supplementary and creative ways, as defined by McGarr (2009), was part of embedding podcasting into the program. McGarr (2009) summarises the issues:

> While podcasting has the potential to enhance the students' learning experience, it can also reinforce the worst aspect of the transmission model of learning. For this reason, future use of these technologies should be learner led, rather than technology led... Future uses of the technology should be guided by sound educational goals that aim to improve the students' existing experience, rather than being guided by vague claims of revolutionizing it.
>
> (p. 319)

Student responses to the iPod initiative, which were accessed through surveys and focus groups, did not concur with McGarr (2009, p. 319). The student responses indicated that podcasting enabled flexible access to course material and provided opportunities for them to upload material of their own creation. However, as we argued above, it was important to design podcasting and the associated technologies so that it was never entirely a one-way transmission. All podcasts contained follow up ideas for personal research and transmission, critical evaluation and reflective discussion, either during the weekly online tutorials or on the server in the open discussion spaces. The combination of the podcast with the mobile device meant that the students were able to control when and how they accessed the podcast – a positive feature noted by students who were time poor and had multiple agendas operating in their home lives.

> I think the use of iPods is fantastic and puts the course into a relevant context for teaching this generation of students. I would recommend more of the 'intensive' sessions are also placed into podcast so we can refer back to them; particularly as we prepare our philosophy statements. Sometimes it is nice to be able to go back over a lecture or tutorial as the topic becomes applicable to a task you are involved with. I personally found the iPod a fantastic tool not only for teaching students, but also so I could manage the load of this course. As a busy mum, with work commitments and family commitments, the iPod has allowed me to listen (and sometime re-listen) to lectures, podcasts etc while driving, watching the kids playing basketball, and the list goes on. The fact that you can have it in your bag, for use anytime, anyplace use, is extremely practical (unlike your computer or an iPad).
>
> (Student comment)

Data from the open comments and focus groups indicate that students were positive about the value of making and listening to a range of short audio products as well. One of the open online comments makes this clear:

> Dear Honest [NAME]
>
> I was a bit like you, I tend to use my laptop and phone for most things. BUT I have started to use my ipod as a one on one teaching tool. I have recorded story books for a Grade 1 student, he enjoyed listening at the same time followed the print and pictures. This adds another dimension to his learning, he is intellectually disabled. I have down loaded a metric conversion app, atlas app and dictionary app; again I use these for one on one teaching, mainly for a Grade 5 student who has learning issues. It appears to me regardless of learning abilities, technology – gadgets really – attract the attention of students, they are not afraid to touch, click, drag. I am hoping to use the ipod during practicum... I'll let you know the progress I make – stay positive.

Overall, the experience of podcasting within the program has demonstrated the capacities for technological tools to be adapted to support learning in line with learning theories and learning design. One student encouraged her own students to bring mobile devices from home (see Walta & Nicholas, 2013). She reflected more general attitudes of the students, who were motivated to use the technology and consider its application to future school contexts. The positive experience of embedding a mobile learning device into the program has been responsible for the adaption of a culture where the program is continuously embracing new technologies to achieve learning outcomes appropriate for teachers who will themselves be required to be creative and adaptable to sustain learning programs in their classrooms.

Overall the use of iPods within the program demonstrated that technological tools could be utilised to demonstrate how effective curriculum design could be underpinned by sustainable technology. It also demonstrated that learning with technology enabled a flexible and creative experience and developed skills which could be adapted to future advancements in software and hardware. Students selecting the program had overwhelmingly endorsed the flexibility of online delivery as their motivation for participation. Exit data was affirming for the introduction of iPods into the program as a tool for improving options for flexibility, creativity and shared engagement in learning.

Principles of design and program sustainability

Teaching and teacher education continue to be confronted with: (1) the dual (and under some circumstances competing) challenges of on the one hand keeping pace with technological developments and on the other hand identifying and then enacting enduring values to shape the processes and outcomes of education; (2) achieving some kind of balance between individual growth and group cohesion; and (3) critical reflection and enculturation into established socio-cultural patterns. For pre-service teacher education programs this can be a particularly profound challenge (see Ng, Nicholas, & Williams, 2009) as they promote and respond to rapid and intense changes in their students' perspectives about themselves, their relationships to their learners and the values that they select to inform their practices.

The growth in technological capacity adds a layer to these decisions that has the potential to increase tensions in a number of different ways. The tensions can be increased by differences in technological knowhow and variation in the associated sense of comfort in working with mobile devices. These tensions can leave some students feeling very frustrated and isolated while others experience freedom and creativity (see Walta & Nicholas, 2013). Technology can challenge in profound ways the teacher's sense of 'control' over what occurs in 'their' class if students use mobile devices to obtain information (or solutions to problems) that go beyond the teacher's expectations. The same devices can challenge trust between teachers and students if devices are used in unexpected ways, e.g. filming the teacher while teaching or other students while learning (and/or posting critical episodes in publicly accessible places). Gaining access to the necessary technology can also be a source of profound concern in 'simple' economic terms, leading teachers to worry about whether the uses of technology are either democratic or equitable.

In each of these situations, there can be profound reasons why pre-service teachers might be reluctant to engage with technology as part of their professional preparation and/or as part of their subsequent practice. It is therefore vitally important that the program of engagement with mobile devices specifically and newer technologies in general is thoughtfully designed and capable of variation in ways that remain consistent with identified desirable principles. Below we indicate the design principles that were followed to enable the mobile learning program to be sustained through many changes in technology and in the overall institutional environment. The changes in technology have been indicated in the description of the program. The changes in the overall institutional environment included changes in reporting arrangements of the program leaders, dramatic increases in student numbers, changes in staffing, changes in overall budget allocations and changes in the perceived distinctiveness of the program within the overall institutional profile.

In response to these challenges, the principles that informed the design of the program and contributed to its sustainability were:

1 Collaborative partnerships that support technological innovation (set up so that learners and staff are all learning through and about technology):

   From the beginning of the program it was clear that technological facility would be unevenly shared among participants and that with a relatively high student/staff ratio and lack of regular face-to-face contact between participants, it would not be possible to provide centralised support to all individuals in their use of the technology. Further, given the range of contexts in which students would be working and for which they would be preparing as teachers, it would not be possible to develop a recipe-like approach to technology in general and use of the mobile devices in particular. Thus, the idea of learning partnerships became a key focus. Not only was this part of the general design of the program, which had a focus on group work and collaboration, but from the very first workshop where the students gained hands-on experience with the devices, emphasis was placed on the students seeking opportunities to work with one another. Explicit room was created for students to step forward as technological leaders. In responding to these opportunities, one of the students created a blog that served as not only a technological but also a pedagogical resource for other students. Other students formed LMS-based groups where they undertook specific project-focused tasks that made use of the devices. Freed of the responsibility for technical support, staff made use of the opportunity to demonstrate a 'having a go' approach with unfamiliar technologies. Students were encouraged to be creative and innovative with technologies and share outputs in the common learning spaces of the Mac OS server.

2 Technological change and innovation associated with improved learning outcomes (that remains the *constant* – not technology for its own sake):

While the initial deployment of the mobile devices took advantage of a new technical capacity, the opportunity to include wireless access in a device that was not a phone, subsequent developments did not prioritise chasing the latest technical innovation. Keeping abreast of technical innovation was not discouraged, but the driver behind this was the attempt to achieve identified pedagogic goals such as more flexible use of images, greater interactivity among learners and also between learners and teachers, or the employment of the devices with diverse groups of learners. So, for example, the program resisted incorporation into the University's general lecture capture technology because of its implementation as a one-way transmission-recording tool. Despite the technical difficulties created, but with sustained support from central technical facilities, the server with its two-way podcast sharing capacity was maintained for longer than originally intended and until general university technical facilities had shifted their orientation to multi-directional interaction. Consistent with that approach, and with assistance from faculty-level educational designers, greater use was made of virtual classroom applications so that all students would have experience of synchronous use of technology.

3 Better access:

More opportunities for collaboration: the overall design of the program reflected collaborative approaches. Small groups were established within the larger cohort of students so that groups of students would have consistent opportunities to interact with one another and that these groups of students would also be able to establish ongoing relationships with particular members of staff and through them with the staff team as a whole. As one aspect of this approach, the program brought in teachers from local schools who were working with similar devices and encouraged students to bring their experiences from schools back to their peers. To support this kind of collaboration, the wiki functionality within the Mac OS server was exploited to provide a means for the students to reflect on their school-based (individual) experiences and staff from the program negotiated with schools for the student's mobile devices to be able to gain access to school-based wireless networks. Similar program features that encouraged greater collaboration between students included sharing of podcasts both made and sourced on the wiki server and sharing useful links to websites and *YouTube* associated with building skills and creative use of technology.

4 More flexible presentation options:

Not only did the use of programs such as *Collaborate* enable students to explore online presentations to their peers (and others), the exploitation of the server in association with podcasting tools was explicitly designed so that the students could make use of their mobile devices to prepare, upload or view student-based material.

5 More flexible communication options:

A feature woven throughout the life of the program has been the use of a diversity of communication tools. For this reason, among others, the program is presented as 'blended learning' and students are made explicitly aware of the need to communicate both electronically and face-to-face. Similarly, staff are accessible via phone and in *Collaborate* sessions. As new options appeared, for example, as *Twitter* became a more popular tool staff started to demonstrate how it could be used to share information.

6 Broader learning opportunities:

Greater exposure to the possibilities of technological devices and software for application to the teaching environment has evolved as part of the program design with the advent of ownership of personal mobile devices and the multiple options for placing personal creative output in cyber spaces which can be accessed by invitation. Assessment options have been broadened to allow for choice of device, software and presentation options. Student output is stored and shared as a point of principle. New devices are introduced with a focus on their potential for creating, supporting, collaborating and accessing material.

Sustainability has underpinned the management of the program from its inception. Fundamentally, sustainability is ensured through building the program on a sound base of research into pedagogical principles about learning through collaboration. Within this context and in the virtual delivery experience afforded by the program, mobile technological devices are essential but evolutionary. The principles of enabling the sharing of knowledge ensure that newly emerging technologies are always embraced for their relevance and affordances. Students now readily link multimedia creations to *YouTube* and other multiuser access sites. Other issues involved in sustainability, such as the variation in student skill levels, the need for technological support, the need for an innovative approach to utilising a range of software in designing learning programs and the cost of constantly upgrading devices, are all addressed within the design of the program, as has been illustrated. The course must be structured in such a way that accessing information, co-constructing knowledge and presenting output are modified to reflect evolving technologies. At the same time fundamental principles of learning, with focus on a virtual environment underpin the overall design of the program.

Long-term evaluation and reflection

What are the enablers and constraints in an online environment seeking to achieve technological sustainability? As we indicated above, the introduction to the program of a mobile technological device, the iPod, was initiated by a specific-purpose funding source. However, the program has outlived both that funding source and the specific devices that the program supported. By not focusing on a single device, but by seeking to connect devices within a suite of technologies, we were able to allow the program to grow in ways that were consistent with our broad principles of education: the creation of a Community of Inquiry, a key role for teachers and teaching and a constructivist, dialogic approach to the organisation of the program.

Notwithstanding program design, remaining technologically current is an ongoing challenge as change to program design may have a ripple effect – new assessment tasks, upgraded software and hardware. BYOD is an almost necessary response, but simply allowing the learners to bring their own devices is not sufficient. There must be a clear sense of purpose for the devices and the community. There must also be consideration of how the differences between initial experiences, attitudes and equipment of the learners can be built into the overall program.

The imperative for creative, flexible thinking in ongoing program design and evaluation for better use of technology will be a challenge for many. However, current and future educators need the same commitment to the basic principles of design and technology-enabled curriculum plus the capacity to negotiate those principles. By designing the program around a sense of community and ensuring that teachers and teaching were key players, we have created a mechanism whereby at least some of these principles have been implemented and sustainability is ensured into the future.

References

Beetham, H., & Sharpe, R. (2007). *Rethinking pedagogy for a digital age: Designing and delivering e-learning*. London: Routledge.

Biggs, J., & Tang, C. (2007). *Teaching for quality learning*. Maidenhead: Open University Press/McGraw Hill.

Brinkerhoff, J., & Koroghlanian, C.M. (2007). Online students' expectations: Enhancing the fit between online students and course design. *Journal of Educational Computing Research*, *36*(4), 383–393.

Brown, R.E., (2001). The process of community building in distance learning classes. *Journal of Asynchronous Learning Networks*, *5*(2), 18–35.

Chickering, A.W., & Gamson, A.F. (1987). *Seven principles for good practice in undergraduate education*. Racine, WI: The Johnson Foundation Inc/Wingspread.

Cochrane, T.D. (2010). Exploring mobile learning success factors. *ALT-Journal of Research in Learning Technology*, *18*(2), 133–148.

Dale, C., & Pymm, J.M. (2009). Podogogy: The iPod as a learning technology. *Active Learning in Higher Education*, *10*(1), 84–96. doi:10.1177/1469787408100197.

Dettori, G., Giannetti, T., & Persico, D. (2006). SRL in online cooperative learning: Implications for pre-service teacher training. *European Journal of Education*, *41*(3/4), 397–414.

Draper, P., & Hitchcock, M. (2006, December). Podcasting: Can the eLearning experience inform eResearch in the digital arts? Paper presented at the CreateWorld 2006 conference and published on the conference website as an enhanced podcast, Griffith University, Brisbane. www29.griffith.edu.au/createworld

Durrington, V.A., Berryhill, A., & Swafford, J. (2006). Strategies for enhancing student interactivity in an online environment. *College Teaching*, *54*(1), 190–193.

Edirisingha, P., Salmon, G., & Fothergill, J. (2006). Profcasting: a pilot study and a model for integrating podcasts into online learning. Paper presented at the Fourth EDEN Research Workshop. Spain: Castelldefels, 25–28. Retrieved on August 2, 2012 from http://www2.le.ac.uk/departments/beyond-distance-research-alliance/projects/impala1/presentations/Berlin/Pilot%20study/view.

Evans, C. (2008). The effectiveness of m-learning in the form of podcast revision lectures in higher education. *Computers & Education*, 50(2), 491–498. Retrieved on December 7, 2014 from http://dx.doi.org/10.1016/j.compedu.2007.09.016

Garrison, D.R. (2011). *E-learning in the 21st century: A framework for research and practice* (2nd Edn). New York, NY: Routledge.

Garrison, D.R., & Arbaugh, J.B. (2007). Researching the community of inquiry framework: Review, issues and future directions. *The Internet and Higher Education*, *10*(3), 157–172.

Garrison, D.R., & Kanuka, H. (2004). Blended learning: Uncovering its transformative potential in higher education. *The Internet and Higher Education*, *7*(2), 95–105.

Garrison, D. R., Anderson, T., & Archer, W. (2000). Critical inquiry in a text-based environment: Computer conferencing in higher education. *The Internet and Higher Education*, *2*(2–3), 87–105.

Goodyear, P. & Dimitriadis, Y. (2013). *In media res*: Reframing design for learning. *Research in Learning Technology*, *21*(Supplement), 19909ff.

Hardman, Frank. (2008). The guided co-construction of knowledge. In N. Hornberger & M. Martin-Jones (Eds.) *Encyclopaedia of language and education* (pp. 253–264). New York, NY: Springer.

Heilesen, S. B. (2010). What is the academic efficacy of podcasting? *Computers & Education*, *55*(3), 1063–1068.

Herrington, J., Mantei, J., Herrington, A., Olney, I.W., & Ferry, B. (2008). New technologies, new pedagogies: Mobile technologies and new ways of teaching and learning. In R. Atkinson & C. McBeath (Eds.), *Annual Conference of the Australasian Society for Computers in Learning in Tertiary Education*, 2008, (pp. 419–427). Melbourne, Australia: Deakin University.

Hiltz, Starr Roxanne, Wellman, Barry. (1997). Asynchronous learning networks as a virtual classroom. *Communications of the ACM*, September 1997, *40*(9), 44–49.

Kearney, M., Schuck, S., Burden, K., & Aubusson, P. (2012). Viewing mobile learning from a pedagogical perspective. *Research in Learning Technology*, *20*(1): 14406ff.

King, K. P. (2002). Identifying success in online teacher education and professional development. *The Internet and Higher Education*, *5*(3), 231–246.

Korthagen, F., Loughran, J., & Russell, T. (2006). Developing fundamental principles for teacher education programs and practices. *Teaching and Teacher Education*, 22(8), 1020–1041.

Kress, G., & Pachler, N. (2007). Thinking about the 'm' in m-learning. In N. Pachler (Ed.), Mobile learning: towards a research agenda. *Occasional Papers in Work-based Learning* 1, WLE Centre for Excellence, London.

Laurillard, D. (2007). Pedagogical forms of mobile learning: Framing research questions. In N. Pachler (Ed.), *Mobile learning: Towards a research agenda* (pp. 33–54). London: WLE Centre, Institute of Education.

Leander, K., Phillips, N. & Headrick Taylor, K. (2010). The changing social spaces of learning: Mapping new mobilities. *Review of Research in Education*, *34*(1), 329–394.

Lee, M. J.W., & Chan, A. (2007a). Pervasive, lifestyle-integrated mobile learning for distance learners: an analysis and unexpected results from a podcast study. *Open Learning: The Journal of Open and Distance Learning*, 22(3), 201–218.

Lee, M. J.W. & Chan, A. (2007b). Reducing the effects of isolation and promoting inclusivity for distance learners through podcasting. *Turkish Online Journal of Distance Education – TOJDE*, 8(1), 85–104.

Lemon, N. (2013). @Twitter is always wondering what's happening: Learning with and through social networks in higher education. In Patrut, B., Patrut, M., & Cmeciu, C. (Eds.). *Social media in higher education: Teaching in Web 2.0* (pp. 237–261). Hershey, Pennsylvania, USA: IGI Global.

Lemon, N. (2015). Confidence to tweet: pre-service teachers engaging with twitter as a professional online learning environment. In Wright, R. (Ed.). *Student-teacher interaction in online learning environments*. Hershey, Pennsylvania, USA: IGI Global.

Lonn, S., & Teasley, S. (2009). Podcasting in higher education: What are the implications for teaching and learning? *The Internet and Higher Education* 12(2), 88–92.

Mayes, T. & Frietas, S. (2007). Learning and e-learning – The role of theory. In H. Beetham & R. Sharpe (Eds), *Rethinking pedagogy for the digital age* (pp. 13–25). London: Routledge.

McGarr, O. (2009). A review of podcasting in higher education: Its influence on the traditional lecture. *Australasian Journal of Educational Technology* 25(3), 309–321. Retrieved on December 7, 2014 from http://ascilite.org.au/ajet/ajet25/mcgarr.pdf

McLoughlin, C. & Lee, M.J.W. (2008). The three P's of pedagogy for the networked society: Personalisation, participation and productivity. *International Journal of Teaching and Learning in Higher Education*, *20*(1), 10–27.

Ng, W. (2012) *Empowering scientific literacy through digital literacy and multiliteracies*. New York, NY: Nova Science Publishers.

Ng, W. & Nicholas, H. (2013). A framework for sustainable mobile learning in schools. *British Journal of Educational Technology*, *44*(5), 695–715.

Ng, W., Nicholas, H., & Williams, A. (2010). School experience influences on pre-service teachers' evolving beliefs about effective teaching. *Teaching and Teacher Education*, *26*(2), 278–289.

Nicholas, H. & Starks, D. (2014). *Language education and applied linguistics: Bridging the two fields*. London: Routledge.

Oliver, R. (1999). Exploring strategies for online teaching and learning. *Distance Education*, *20*(2), 240–254.

Pachler, N., Bachmair, B., & Cook, J. (2010). *Mobile learning: Structures, agency, practices*. New York, NY: Springer.

Palloff, R.M., & Pratt, K. (2003). *The virtual student. A profile and guide to working with online learners*. San Francisco CA: Jossey-Bass.

Pawan, F., Paulus, T., Yalsin, S., & Chang, C. (2003). Online learning: Patterns of engagement and interaction among in-service teachers. *Language Learning & Technology*, *7*(3), 119–140.

Ragusa, A. T., Chan, A., & Crampton, A. (2009). iPods aren't just for tunes: exploring podcasting as a socio-cultural technology facilitating student experiences of higher education in rural Australia. *Information, Communication & Society*, *12*(5), 678–690. Retrieved on December 7, 2014 from http://libezproxy.open.ac.uk/login?url=http://search.ebscohost.com.libezproxy.open.ac.uk/login.aspx?direct=true&db=ufh&AN=43744012&site=eds-live&scope=site

Rovai, Alfred. (2002). Building a sense of community at a distance. *International Review of Research in Open and Distance Learning*, *3*(1), no page numbers. Retrieved on March 13, 2015 from http://www.irrodl.org/index.php/irrodl/article/view/79

Salmon, G., & Edirisingha, P. (Eds) (2008). *Podcasting for learning in universities*. Maidenhead: Open University Press.

Shackelford, J.L., & Maxwell, M. (2012). Sense of community in graduate online education: Contribution of learner to learner interaction. *The International Review of Research in Open and Distance Learning*, *13*(4), no page numbers. Retrieved on March 13, 2015 from http://www.irrodl.org/index.php/irrodl/article/view/1339/2317

Sharples, M., Taylor, J., & Vavoula, G. (2007). A theory of learning for the mobile age. In R. Andrews and C. Haythornthwaite (Eds.) *The Sage handbook of elearning research* (pp. 241–247). London: Sage.

Shea, P., Li, C., & Pickett, A. (2006). A study of teaching presence and student sense of learning community in fully online and web-enhanced college courses. *The Internet and Higher Education*, *9*(3), 175–190.

Sheridan, K., & Kelly, M. (2010). The indicators of instructor present that are important to students in online courses. *MERLOT Journal of Online Teaching and Learning*, *6*(4), no page numbers. Retrieved on March 13, 2015 from http://jolt.merlot.org/vol6no4/sheridan_1210.htm

Stephenson, J. (Ed.) (2001). *Teaching and learning online pedagogies for the new technologies*. London: Kogan Page.

Swan, K.P., Richardson, J.C., Ice, P., Garrison, D.R., Cleveland-Innes, M., & Arbaugh, J.B. (2008). Validating a measurement tool of presence in online communities of inquiry. *Ementor*, 2(24). Retrieved on December 7, 2014 from http://www.ementor.edu.pl/artykul_v2.php?numer=24&id=543

Walta, C., & Nicholas, H. (2013) The iPod Touch in association with other technologies in support of a community of inquiry in off-campus teacher education. *Australasian Journal of Educational Technology*, 29(6), 870–886.

9 Sustaining mobile learning with pervasive game

An example of cultural history exploration

Ju-Ling Shih, Jyun-Fong Guo and Cheng-ping Chen

Introduction

In recent years, the advancement of digital technology has changed the practice of human society. The advancement of computer technology has attracted intensive and crucial discussions and debates on the issues of "learning to use technology" and "learning with technology" in school education. In the realm of communications technology, digital convergence has integrated the web, digital platform, Web 2.0 and cloud technology, have turned the one-way transmissive mode of teaching to collaborative knowledge construction and sharing. Advancements in information technology beyond the year 2000 have pushed the instructional applications of computer technology to go further to break through the limitations of time and space to extend the classroom beyond its physical constraints. The innovation of mobile and interactive technology has made communication in the ubiquitous context possible. The terms "mobile learning (mlearning)" and "ubiquitous learning (u-learning)" have become more well-known to the general public. The mobile device is subject to no constraints in time and space. Laine et al. (2010) defined mlearning as a form of informal learning where the learner traverses a physical context or contexts carrying a personal mobile device which provides learning materials and activities. Mikic et al. (2007) believed that mobile learning has created a new learning environment that enables learners to learn at any time and anywhere, through action learning aids, teaching materials, teachers and other learners.

The 2009 Horizon Report has predicted mobile learning to be an important development for digital learning (Johnson et al., 2009) and the 2011 Horizon Report predicted that more than 80% of people will use mobile devices connected to the Internet in 2015. More importantly, by 2016, mobile device use in education would outnumber the use of computers (Johnson et al., 2011) in the classroom. With the help of advanced mobile devices and well-designed instructional strategies, mlearning is enabling students to walk out of classrooms and into real life scenarios to interact with real learning environments. Studies (e.g. Laine et al., 2010) have also showed that mlearning and u-learning are especially helpful in facilitating learners' construction of knowledge in regard to the situational information in social, cultural and physical contexts.

Well-designed instructional strategies are essential for successful mlearning. However, many local studies on mlearning simply reiterated the concerns of establishing a learning system and emphasising the use of the mobile devices (such as PDA) with the established system (e.g. Liang, 2005; Lin, 2006). These studies appear to fall short of designing adequate strategies to carry out effective instruction. The system-design type of studies may have caused instructors, especially high school teachers, to be reluctant to adopt mobile technologies due to the lack of budget support and knowledge of applicable instructional strategies. Therefore, Ng and Nicholas (2013) proposed a person-centred model indicating "good communication and support, being consultative and maintaining trust between management, teachers, students and technical support personnel" (p. 4) as the core actions for implementing mobile learning.

Other than the instructional issues, making mobile learning sustainable is the current issue. Cisler (n.d.) discussed the sustainability issue of information and communication technology (ICT) in education that encompassed four dimensions, namely economic sustainability, social sustainability, political sustainability, and technological sustainability. It became important to seek a mechanism that would eliminate technological implementation costs, and intrigue users to choose knowledge inquiries as individuals and as groups. As a result, in this research, a pervasive game was used to design mobile learning using participants' self-owned mobile phones with Internet connections, their personal social network service accounts and leisure time; therefore little technical and instructional efforts were required to implement this inquiry-based learning strategy.

An innovative way to accomplish mlearning using the instructional pervasive gaming (IPG) approach has been introduced by various educational researchers in the past ten years (Laine et al., 2010). Montola (2011) explained a pervasive gaming environment as "an extension of mlearning with an emphasis on the roles of an intelligent environment and of the context". The advancement of mobile technology has brought about not only the booming of m-business but also the imaginary space of mobile games (Chen, Guo, & Shih, 2012). Okazaki, Skapa, and Grande (2008) recognised that mobile games have quickly opened up new dimensions for entertainment applications and have rapidly become an attractive alternative to PC-based games.

In mobile learning strategies, game-based mobile learning has been recognised as effective (Huizenga et al., 2009; Schwabe & Göth, 2005). Related research (Facer et al., 2004) also demonstrated that pervasive gaming has significant effects on learning. Recent research (Schwabe & Göth, 2005; Shih & Chen, 2012) indicated that mobile gaming has positive user effects, especially in educational applications. Schwabe and Göth (2005) designed an instructional system that combined mobile learning and mobile gaming. This instruction used college students as learners who were divided into groups to play the game. They obtained information from the website regarding the mission and geographical coordinates of the locations. They then used GPS to complete a task and uploaded the result back to the workstation before they could get the location information of the next mission. The winner was the group who completed the task first.

In mlearning, the learning of content is location-sensitive as learners are free to move around. The advantage of employing mobile technology in learning is not the mobile technology itself, but the unique attributes of the mobile devices that provide an environment upon which one can apply innovative learning theories and learning strategies (Su, 2005). With the aid of GPS, Chen and Shih (2011a; 2011b) initiated a study employing a typical pervasive game, geocaching, to teach high school students geographical coordinates and map reading. The results indicated that GPS-aided geocaching helped students to learn to read maps and recognise the geographical coordinates in a more effective way. In addition, the GPS-aided geocaching group generally acquired a better attitude toward the use of technology.

Such results are in accordance with some other related studies (Chavez, Schneider, & Powell, 2004; Mandy, 2006). However, Gentes, Guyot-Mbodji and Demeure (2010) pointed out that most game strategy does not take into consideration the anthropological data pertaining to the specific context of use. They argued that the advancement of geolocalisation helps us to concentrate on the diversity of location-sensitive experiences rather than focusing uniquely on the geographical skills of players engaged in a treasure hunt pursued by other players in their game design. They further indicated that maps and trajectories are one aspect of treasure hunts, while Shih and Chen (2012) indicated that companionship and discovery are at the core of the pleasure of pervasive games. Walther (2005) recognised that a mobile game is one that takes into account the changing of relative or absolute position/location in the game rules. This excludes games for which mobile devices merely provide a delivery channel where key features of mobility are not relevant to the game mechanics. Corey (2010) stated that both urban and environmental histories in particular lend themselves to hands-on approaches that link the classroom with tangible examples of historical events and trends through the pedagogical design of the learning of place. Huizenga et al. (2009) also claimed that mobile and location-based technologies provided opportunities to embed learning in authentic environments, thereby promoting enhanced engagement and learning outside of traditional formal educational settings. They conducted a quasi-experiment where the students from 10 classes played a mobile history game while the students from another 10 classes received a regular, project-based lesson. The results showed that those students who played the game were more engaged and gained significantly more knowledge about medieval Amsterdam than those students who received regular project-based instruction.

The goal of this research was to establish a pervasive game framework to design an exploratory game that allows participants to explore historical locations to learn about interrelated knowledge. The historic site, Provintia, in Tainan, was chosen to be the gaming location. Its related history and cultural content were embedded in the multimedia form with augmented reality (AR) interactions. The participants used the social networking service, *Facebook*, to document group activity, share information, and interact with group members. The implications for sustaining pervasive gaming were explored and discussed.

Literature review

Pervasive games

A pervasive game (PG) is a mobile game that seamlessly integrates the physical environment with virtual scenarios. According to Walther (2005), a PG implies a mixed-reality gaming space; that is, the gameplay connects the virtual world with the tangible world. He further indicated that the mobile mechanisms (that include software, hardware and information), should be in place throughout the gameplay to facilitate and ensure that there are fluent communication channels between the virtual and physical worlds. Pervasive games often blur the boundary between games and real life.

Laine et al. (2010) commented that PG is an extension of mlearning. The interaction between the player and the environment is the key point. Another key point about PG was raised by Montola (2011). He stated that the "magic circle of play" created by an online game is an invisible fence keeping the game from real life. He further explained that the magic circle of play is a social and cultural contract that separates ordinary life from play, where players are also separated. However, PG breaks the circle and bridges real and virtual environments to accelerate more interactions.

From an educational view point, Laine et al. (2010) described a technology integration model for game-based pervasive learning systems, stating that an instructionally constructed PG needs to meet the following requirements: (1) pedagogical requirements that include user profile, interaction and collaboration, ownership, authenticity and relevance, as well as provision of support and assessment; (2) game design requirements that consider resources (both financial and human), cultural issues, technical issues, environmental issues, social issues and temporal issues; and (3) context requirements that include context-awareness, dynamics, interaction and content.

There are various papers that discuss the design issues of pervasive games. From the perspective of technology applications, Walther (2005) constructed a model and identified the "four axes of PGs". He illustrated the four axes as: (1) distribution: this refers to the effective wide distribution network of the gaming information; (2) mobility: this refers to user mobility, computing mobility, network mobility, and context-aware and cross-platform services; (3) persistence: this refers to the total availability all along; and (4) transmediality: this refers to a media circle that multi-links the world of virtual social networks. He further depicted three critical constructional framework elements of PG: (1) game rules: all PGs must be rule-based and clearly defined; (2) game entities: the abstract class of an object that can be moved and drawn over a game map, which can further be shaped into three categories such as game object, human agent, and physical object; (3) game mechanics: an input-output engine that monitors and modifies the physical and virtual linkage to ensure a fluent game flow.

In PG, players are typically equipped with handheld or mobile devices. WiFi or 3G connections are common technologies used for the players to communicate with the virtual environment. To be more precise, location tracking and orientation sensors are usually based on GPS. AR is also used to increase the gamification of gameplay. According to Benford et al. (2005) pervasive games need three kinds of technology support: (1) displays that can make digital content, which can transmit digital content to users' mobile devices, such as smartphones, PDAs, and tablet PCs; (2) wireless communications, which can connect users with the server and other users by 3G, WiFi, GPRS, and GSM; (3) sensing technologies, which allow the context-awareness mechanism to work for the environment, such as GPS, microphone, camera, other sensors. These mobile technologies allow users to experience the game activities outdoors.

As mentioned above, Gentes, Guyot-Mbodji and Demeure (2010) specifically pointed out that recognition of local culture must accompany the co-creation of content and team exploration. It also strengthens the importance of cooperative learning in pervasive game. According to Pragnell, Roselli and Rossano (2006), cooperative learning is achieved by dividing learners into small groups to accomplish the best results by means of mutual assistance among the group members. This fosters both a team spirit and interpersonal relationship with positive personal effectiveness.

There are many collaborative and cooperative learning studies, but only a few have been discussed in the gaming context. Dyson, Griffin and Hastie (2004) indicated that students can compete more effectively working as a team against another team in a game unit, while Huang, Shih and Lai (2011) studied a cooperative game used to increase the learning effectiveness of an engineering class, and the result was that students in the cooperative learning class significantly outperformed students in other classes. Students' feedback also showed that cooperative learning effectively increased their motivation to learn. The results of these collaborative or cooperative game studies support Gentes, Guyot-Mbodji and Demeure's

(2010) and Edwards' (2003) claim that team cooperation, collaboration, and competition are essential for gameplay.

Augmented Reality

Augmented Reality (AR), also known as Extended Reality, is a computer graphics interactive technology originating from Virtual Reality (VR). Azuma (1997) believed that interactive technology could be applied to military, medicine, industrial design, maintenance, commercial activity, learning and entertainment. Therefore, in recent times, AR has often been used to enhance interaction in the pervasive game and mobile learning activities. So far, AR content recognition techniques include three main types: (1) use of specific images as recognition objects; (2) use of real objects as image recognition targets; (3) use of GPS positioning and digital compass to identify images. As Ultra Mobile PC (UMPC) image processing technique has improved, more and more studies have implemented AR into their mobile learning practices.

AR seems to be essential to pervasive games or mobile edutainment applications. For instance, Cheok et al. (2004) brought the classical arcade game *Pacman* into the real world and called it *Human Pacman*. Players collected cookies made by AR and avoided being caught by the ghosts. The ghosts in the game are avatars representing other players, so it is also a multiplayer game. Other games, *ARQuake* (Piekarski & Thomas, 2002) and *Treasure* (Barkhuus et al., 2005), also used the application of AR to allow the players to collect objects.

Herbst et al. (2008) presented an interactive time travel game, *TimeWarp*, which guides the users to explore historic buildings with the AR virtual tour guide, *Heinzelmännchen*. Participants explored the two-thousand-year history of the city of Cologne, and experienced Medieval Roman times, the new century, and even future time travel. AR in this instance functions not only as a simple game tool, but also as the mobile guide. Ballagas, Kuntze and Walz (2008) designed *REXplorer* as a historical exploration guide for tourists in Regensburg. Tourists visited landmarks and found virtual magical spirits and treasures using detectors on their way.

Billinghurst et al. (2003) claimed that AR is useful to education because using AR as teaching materials can allow students to be immersed in the learning of content of both the virtual and real environment. Students can even succeed in such learning environments without computer experience. In the past decade, the increased simplicity of using AR has brought students closer to the virtual scene as well as the real environment. Perceptual feedback creates fun learning and stimulates students' learning motivation and enthusiasm. AR was used in several educational situations, such as geometry teaching, science experiments (Newton's Law), and nuclear emergency practice (Enyedy et al., 2012; Kaufmann, 2006). The research results suggest that AR can increase students' understanding in the learning of content.

Social networking services

Traditional digital learning (e-learning) has mostly taken the form of using a learning management system. Course content was totally decided by the instructors. Following the development of Web 2.0, the control rights of digital learning have been transferred to the students. A typical Web 2.0 tool is *Wikipedia*, in which people edit the digital content together and learn independently through the process (O'Hear, 2006). Today, the most used Web 2.0 application is the social networking service *Facebook*. Instead of using social networking as a learning platform Liu (2010) used it as a teaching tool to increase students' learning effectiveness.

Mazman and Usluel (2010) believed that social networking services could be used as a teaching channel to extend classroom learning. At the same time, students can interact, discuss, and share information on it, which further helps to develop critical thinking ability. The results of their study suggested that *Facebook* could be effectively used in education. OnlineCollege.org (2009) listed 100 ways of *Facebook* teaching, illustrating the benefits *Facebook* can bring to sustain group communication. In pervasive gaming research, *Facebook* is used for group cooperation, information management, and sharing.

Research methods

The research process was divided into four stages (see Figure 9.1). The first stage was the learning content preparation. The game script was designed around Tainan historical events that happened in the period of Koxinga, and also contained some folklore stories to increase the fun. The activity space was a Grade I national historic monument, Provintia. Provintia was built by the Dutch in 1653 as a trade and administrative centre. Today, the buildings are in the Chinese architectural style of the Ching dynasty because the original fort was damaged a long time ago and became the building foundation of today's Wen Chang pavilion and Temple of Sea God. These buildings have been through times of Dutch, Kingdom of Tungning, Ching Dynasty and Japanese colonial rule, until the Taiwanese restoration. It is the epitome of history with various cultural artefacts and stories.

The learning content of the game was collected through historical literature, books, textbooks, web information, Tainan government tourist bureau information, field study, tour guides, and oral history interviews. The learning content includes four aspects: (1) a historical aspect that includes stories about the fort, ordnance depot, hospitals, and schools in various historical times; (2) a geographical aspect that is largely about the importance of the location and position of the fort during wartime; (3) an architectural aspect that includes objects and features such as stone lions and sculptures; and (4) a cultural aspect that includes historical monuments such as tombstones and wells. In the pervasive game, 14 riddles were

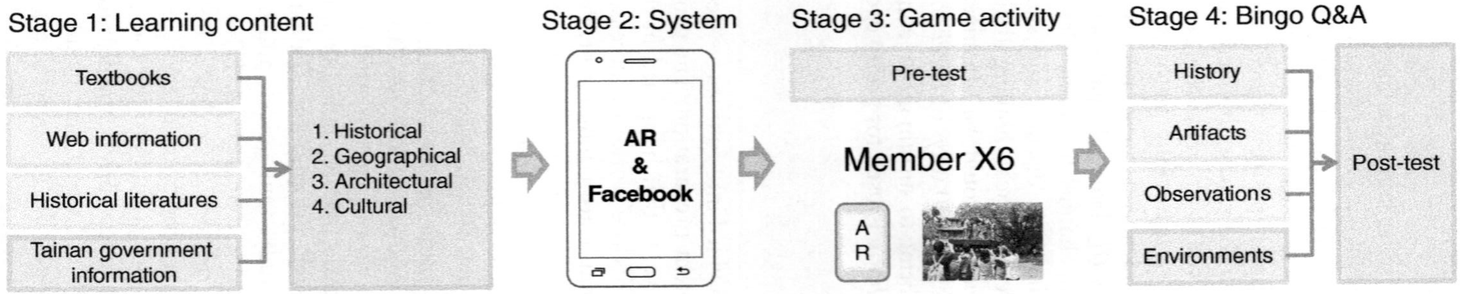

Figure 9.1 The research process

created for the exploration. The same questions were used in both the pre-test and post-test of learning effectiveness.

The second stage was system design. Smartphones equipped with wireless Internet connections were used as the mobile learning tools. The phones were installed with AURASMA (Figure 9.2). AR presentations on the players' mobile phones can increase players' gaming motivations and interests. With the instructional pervasive game design, players can have more interactions with the environment. *Facebook* was installed to allow participants to communicate and share answers with group members. It was not only a communication bridge to players, but also a terminal for researchers where they could oversee the participants' interactions and monitor players' gaming conditions. The game mechanism was implemented to increase group competition and cooperation.

The third stage was the pervasive game activity (Figure 9.3). Thirty college students were invited to participate in the study. They were divided into groups of five with a total of six groups. Every group was accompanied by one research observer whose responsibility was to offer task guidance and solve hardware or software problems. Before the game, *Facebook* group spaces were installed for each group. Then, a pre-test was conducted to gather data on the players' prior knowledge about history. At the same time, a pervasive game effectiveness evaluation pre-test was also conducted to gather information on participants' expectations of the game.

When the participants entered Provintia, tourist brochures were distributed. The pervasive game started with riddle solving. Participants explored the historical spots or cultural objects according to the hints given by the riddles, trying to understand their contents and meanings. The participants used smartphones or tablet PCs to detect the target spots or objects and drive

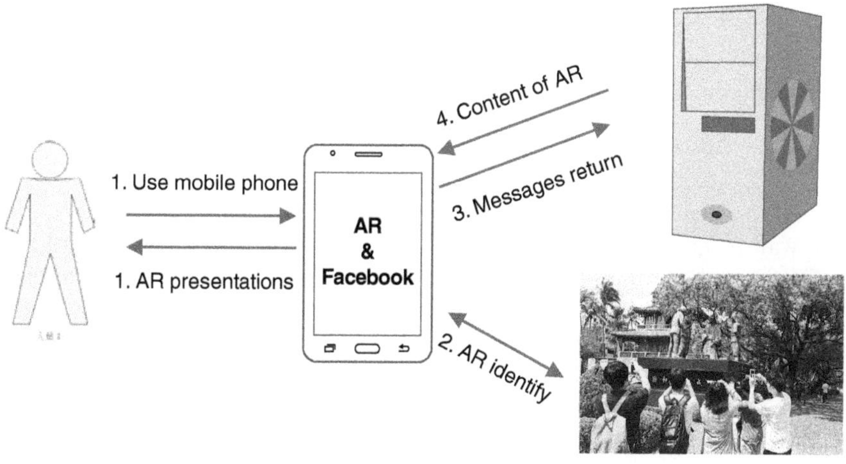

Figure 9.2 Aurasma system data flow

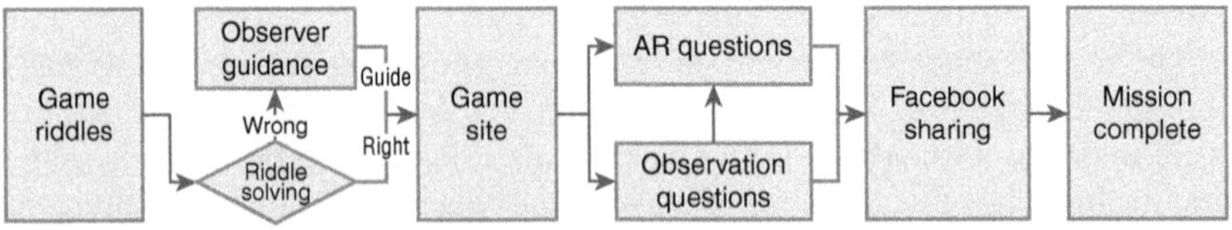

Figure 9.3 Pervasive game process

Figure 9.4 AR content

the AR content to show short clips of historical events or talking figures on the wall to give hints (Figure 9.4). The AR content would repeat until the participants had found the answers to the riddles, documented their findings, and uploaded photos or answers to the *Facebook* community.

After the pervasive game exploration, a group bingo competition of Q&A activity was conducted. The groups took turns to answer questions. There were 49 questions concerning history, artefacts, observations, and environments. They were allowed to retrieve information or answers from the Facebook communications. The purpose of this activity was to get all of the participants together, and refresh their memories of the one-day activity. It was also important for the instructor to call a close to the activity.

The final stage was the data analysis. In the gaming process, all interactions were recorded. All learning paths and the learning content entered during the process were recorded on the mobile devices. Research observers also recorded behaviour observations. After the game, historical knowledge post-tests were conducted. A pervasive game evaluation questionnaire and focus group interviews were also conducted to investigate the effectiveness of the game. They included four aspects, namely game rules, game mechanism, game interaction, and game content. Behaviour observations and interaction records were used for cross-analysis of the evaluations at the end.

Research results

In terms of learning effectiveness, the results of the pre-test (m=31.33) and post-test (m=88.50) showed that the participants' understanding of Provintia history and culture had significant improvements ($t=22.46$, $p=.000<.001$) from the pervasive game (Table 9.1).

Upon analysis of the four parts of the questions, it was found that participants performed better on geographical aspect and architectural aspects, 93.3% and 92.3% respectively on the correction rates; and less well on the historical aspect and cultural aspect, 89.4% and 82.7% respectively on the correction rates. From the student interviews, out of the four types of questions, e.g. history, artefacts, observations, and environments, it was deducted that students performed better on the observational questions because the learning content was more tangible and the related concepts were easier to grasp. Even though the historical and cultural content were more difficult since they were more conceptual and abstract, the AR content provided the participants with cognitive support since the short clips would provide clues to the users. At the same time, it had stimulated students' learning motivations to continue the exploration.

In terms of the pervasive game's effectiveness, four aspects including game rules, game mechanism, game interaction, and game content, were evaluated. Since the pervasive game's effectiveness evaluation pre-test was conducted to gather information on participants' expectations of the game, the post-test was conducted to show how the pervasive game compared to

Table 9.1 T-test results of learning effectiveness of pervasive game

Item	*pre- and post-test*	*n*	*mean*	*SD*	*t*
Provintia history and culture understanding	Pre-test	30	31.33	12.65	22.46*
	Post-test	30	88.50	7.44	

*p < .001

Table 9.2 T-test results of pervasive game activity effectiveness

Item	*pre- and post-test*	*n*	*mean*	*SD*	*t*
Game Rules	Pre-test	30	25.36	2.85	–.803
	Post-test	30	25.86	2.87	
Game Mechanism	Pre-test	30	23.73	3.00	.22
	Post-test	30	23.56	3.21	
Game Interaction	Pre-test	30	29.30	3.09	.100
	Post-test	30	29.23	3.43	
Game Content	Pre-test	30	12.90	3.97	–9.950*
	Post-test	30	20.96	2.35	

*p < .001

the participants' expectations. The t-test results (Table 9.2) show that the pervasive game design was very close to the participants' expectations although some of the results between pre- and post-tests did not reach significant differences.

The game rule aspect investigated whether the challenges, competitions, and feedback would influence the participants' gaming behaviours and attitudes. It included five questions such as "I thought the game rules were clear"; "The bingo game had stimulated me to be more actively involved in the riddle solving process"; and "Solving riddles had given me a great sense of achievement". The pre-test (m=25.36) and post-test (m=25.86) had no significant difference (t=-.803, p=.429>.05). It showed that the game was about the same as the participants' expectation. The participants said in the interviews that they felt excited and happy when able to solve the riddles, which also enhanced their sense of achievement.

The game mechanism aspect investigated whether the use of smartphone, AR, and *Facebook* would influence the participants' gaming behaviours and attitudes. It included five questions such as "AR storytelling had made the site exploration more interesting"; "AR storytelling made me have deeper impression to the historical site"; and "I would read the materials on the *Facebook* posted by other members to gain more understanding to the historical site". The pre-test (m=23.73) and post-test (m=23.56) showed no significant difference (t=.228, p=.821>.05), and the post-test result was lower than the pre-test. From the interviews, participants revealed that they still preferred shorthand note-taking instead of typing into the smartphone so the game mechanism was not as convenient as they expected. Therefore, the contents on *Facebook* were mostly photos or short notes. To improve this in the future, it might be useful to design a quick upload mechanism, keywords, or icons to enhance the participants' motivation in using *Facebook*. Nevertheless, some groups used the *Facebook* communication records to correctly answer the bingo questions at the end of the pervasive game. It showed that *Facebook* usage was quite helpful in some way.

The game interaction aspect investigated the participants' interactions with others and the environment. It included six questions such as "I found information that was not stated on the tourist guide"; "I would cooperate and share works with group members during the gaming process"; and "I like the AR contents more than the textual descriptions on the tourist signs". The pre-test (m=29.30) and post-test (m=29.23) had no significant difference (t=.100, p=.921>.05), and the post-test result was slightly lower than the pre-test. From the interviews, participants pointed out the hardware factors had influenced their motivation to continue the game with enthusiasm. Sometimes AR image recognition failed due to sunlight, shadows, or photo taking angles. Therefore, a rich image database might be necessary to increase the interaction fluency.

The game content aspect investigated the participants' meta-cognition of the game. It included four questions such as "I have more understanding of the historical background of the site"; "I have more understanding of the architectural features of the site"; and "I have more understanding of the cultural artefacts of the site". The pre-test (m=12.90) and post-test (m=20.96) had reached significant difference (t=-9.950, p=.000<.001). After playing the pervasive game, participants generally had more understanding of Provintia, and could even tell the stories to others.

An IPG framework

PG is a relatively new term for most instructional gaming researchers and developers. The idea of PG started at the beginning of this century. Even though numerous mixed-reality game prototypes have been built and studied (Montola, 2011), a comprehensive model and guidelines for developing instructional PG are still being developed.

Gentes, Guyot-Mbodji and Demeure (2010), recognised local culture to be a part of pervasive gaming, especially with Instructional Pervasive Game (IPG). They suggested that for a cultural-rich PG, three key features should be included: (1) collaborative content where the content should be designed by the local community – people with qualified knowledge or who are actually living on the premises, therefore, scholars, neighbours, and respectable people alike are all contributors of the content; (2) team exploration where the gameplay either relies on solitary errands or on collective sharing involving strategies that can be more "group oriented"; and (3) cultural narrative where there is always a first narrative that sufficiently describes the culture of the city to provide game players with a clear view of the fashionable places, the living or working areas, and the cultural spots.

A review of relevant literature and IPG implementation experiences helped in the construction of a theoretical framework. The framework defines that a cultural-inquiry PG should accommodate technology convenience, cultural richness, and game fluency. A set of criteria was also defined to verify the game flow, the game roles and contexts, and game collaboration. A conceptual

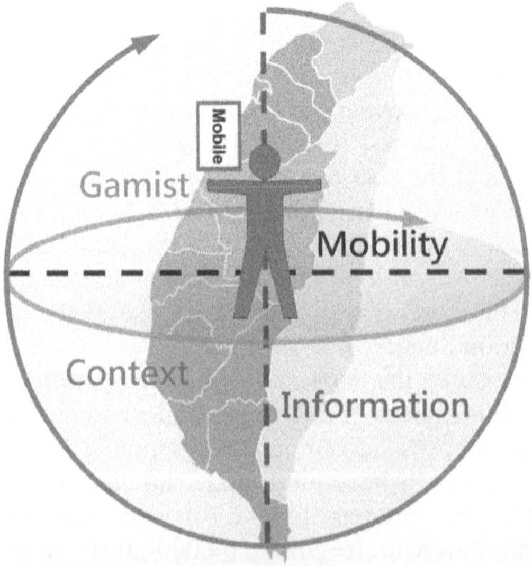

Figure 9.5 Instructional Pervasive Game learning model

model was formed in relation to the framework (Chen & Shih, 2012) (see Figure 9.5). The six factors of IPG included the user, environment, information, context, mobility and gamist.

The first two factors, user and environment, are the tangible factors in the model. The user is the student who is the central learning subject. Different from lecture-type instruction where the teacher is the centre of the learning process, students play the main role in the learning process in IPG. The learning environment refers to the physical environment, which is not confined to the classroom but the whole living environment where the students can physically reach and experience learning. In our case, Taiwan is the main location where learning happens.

The third and the fourth factors are the information and the context. The information refers to the teaching materials that are selected, designed, and organised by content experts (in most cases these are the teachers) which are taught both in class, presented on the mobile devices, and referred to in the physical environment and explored by the students. As the information given by the teachers seems to be linearly structured material, it guides the students to extend their learning spaces to the environment that forms a bigger space that is the learning context. As shown in the figure, the information axis leads the students to "spin" around the learning space that forms the surface of context. The learning context is the abstract and conceptual part of the learning, which acts as the horizon of the learning ground that is expanded into a surface. It includes both the mechanical data and the anthropological data that are experienced, collected, and formed in the exploration process. It is also the process when the students are constructing their own knowledge. What's important is that the anthropological data are those qualitative raw materials collected by the students which include photos, interviews, field notes, and so forth. In other words, these two factors play the role of content and context.

The fifth and sixth factors are the mobility and gamist. Mobility refers to user's ability to move around in the learning environment. Other than sitting in the classroom, users not only have the freedom but also the capability to travel and explore with learning content embedded in the mobile devices. As the mobility is the horizontal axis that extends users' learning experiences to the outside of confined classrooms, gamist is the drive that enables the knowledge surface to expand. Gamist, the game rules that lead to cooperation or competition, is the driving force that creates the challenge and adventure scenario. Using a metaphorical description, if mobility is the vehicle that takes the users to places, gamist is the fuel that provides the energy and motivation to surf. The two factors create the surface consisting of motion and motivation.

The six factors together form the learning sphere. The whole sphere can be regarded as the knowledge body that is and is to be gained by the users. The IPG model reflects the Game Flow Model initiated by Sweetser and Wyeth (2005) that concentration, challenge, player skills, control, clear goals, feedback, immersion, and social interaction are key factors that may affect the effectiveness of a computer game. With the addition of the Pervasive Game Flow Model proposed by Jegers (2009), we can show 10 additional key factors for a PG are important to be integrated when implementing IPG:

- Support the players in the process of switching concentration between in-game tasks and surrounding factors of importance.
- Stimulate and support the players in their own creation of game scenarios and pacing.
- Help the players in keeping a balance in the creation of paths and developments in the game world.
- Be very flexible and enable the players' skills to be developed at a pace set by the players.
- Enable the players to pick up gameplay easily.
- Support the players in forming and communicating their own intermediate goals.
- Not imply or require players' actions that might result in a violation of social norms.
- Enable the players to shift focus between the virtual and the physical parts of the game world.
- Enable meaningful and purposeful social interaction within the gaming system.
- Motivate the players to communicate and interact socially.

Conclusion

Since the early 1990s, active or inquiry-based education has received a great deal of attention. This active learning method has opened a new way for social science learning. The National Council for the Social Studies (1994) believed that students should "systematically employ processes of critical historical inquiry to reconstruct and reinterpret the past, such as using a variety of sources and checking their credibility, validating and weighing evidence for claims, and searching for causality" (p. 40). However, Molebash (2004) found that unguided online historical inquiry did not guarantee meaningful learning. Appropriate guidance is necessary for successful historical inquiry-based learning. Manlove, Lazonder, and de Jong (2009) asserted that supporting tools that incorporated goal-lists, hints, prompts, cues and templates reinforced the cognitive regulation of skills of students during a fluid dynamics task. According to their assertions, it could be reasonably assumed that their innovative pervasive game is an appropriate "supporting tool" for benefiting players' information acquisition.

The goal of this research is to establish a pervasive game framework to design an exploratory game that allows participants to explore historical locations to learn about interrelated knowledge. From the trial of this study, suggestions about the use of a digital system would include the implementation mechanism of AR and *Facebook*. The use of markless AR can create a seamless learning environment combining virtual and real environments, and the use of *Facebook* can create group interaction and cooperation. Whether the t-test results between pre-test and post-test had reached significant differences, the scores between tests were close to each other. It meant that the effectiveness of pervasive game was close to the participants' expectations. It showed that pervasive game not only can effectively help the participants understand the history, geography, and culture of Provintia, but also that it can allow the immersive learning mechanism to increase participants' learning motivation.

The main theme of this chapter is to search for a sustainable model of mobile learning that doesn't require high-level technical knowledge. Since smartphones are prevalent, related applications (apps) are also easy and free to download, and mobile Internet connections are common, mobile devices are indeed more generally used for learning purposes. For the purpose of making pervasive game widespread and knowing its sustainability, reaching participants besides youngsters who might not be as familiar with digital technology might be a concern. Once the mechanism is built, a city tour game or countrywide activity might also be applicable.

As this study came to a conclusion, Google Niantic Labs (2013) published a free Android App game, *Ingress*, which is a real-time multiplayer location-based game. Users check into places using the GPS function of their smartphone and interact with the portals established around public arts or landmarks. Viewing alternate reality generated by AR technology, users participate in the making of an open narrative of science fiction. Countless users logged in to the game, which manifested the milestone of the transformation of traditional individual-based inquiry-based mobile learning into an interactive cooperative pervasive game. It also evidences the potential of what was discussed in this chapter concerning the fulfilment of technology convenience, culture-enriched, location-based, context-aware, and team-oriented mobile learning. It is economically, socially, politically, and technologically sustainable (Cisler, n.d.) since pervasive games are self-sustained by nature.

It is our hope that the game proposed in this study can be expanded into a wider area in the future to include more culture and history of Southern Taiwan into the instructional pervasive game, through which formal education materials can be brought more alive for the students.

Acknowledgements

This research was supported by the National Science Council in Taiwan under projects NSC 98-2511-S-024-006-MY2, NSC 100-2628-S-024-002-MY3, NSC 100-2631-S-011-003.

References

Azuma, R. T. (1997). A survey of augmented reality. *Presence-Teleoperators and Virtual Environments*, *6*(4), 355–385.

Ballagas, R., Kuntze, A., & Walz, S. P. (2008). Gaming tourism: Lessons from evaluating Rexplorer, a pervasive game for tourists. *Pervasive Computing, LNCS 5013*, 244–261.

Barkhuus, L., Chalmers, M., Tennent, P., Hall, M., Bell, M., Sherwood, S., & Brown, B. (2005). Picking pockets on the lawn: the development of tactics and strategies in a mobile game. In *UbiComp 2005: Ubiquitous computing*. Symposium conducted at the meeting of Springer, Berlin.

Benford, S., Rowland, D., Flintham, M., Drozd, A., Hull, R., Reid, J., Morrison, J., & Facer, K. (2005). Life on the edge: Supporting collaboration in location-based experiences. In *Proceedings of the SIGCHI Conference on hHuman Factors in Computing Systems*. Symposium conducted at the meeting of ACM, New York.

Billinghurst, M., Belcher, D., Gupta, A., & Kiyokawa, K. (2003). Communication behaviors in colocated collaborative AR interfaces. *Computer Interaction 16*(3), 395–423.

Chavez, D. J., Schneider, I., & Powell, T. (2004). The social-psychology of a technology driven outdoor trend: Geocaching in the USA. *Hawaii International Conference on Social Sciences*. Symposium conducted at the meeting of Honolulu, Hawaii.

Chen, C. P., & Shih, J.-L. (2011a). Learning geographical coordinates and map reading with GPS-aided geocaching. *The 10th World Conference on Mobile and Contextual Learning*. Symposium conducted at the meeting of Beijing Normal University, Beijing.

Chen, C. P., & Shih, J.-L. (2011b). Enhancing learning effects of map reading and coordinates with GPS-aided geocaching learning activities. *International Mobile Learning Conference 2011 (mLearn 2011)*. Symposium conducted at the meeting of International Ballistics Society, and NDIA, Tokyo.

Chen, C. P., & Shih, J.-L. (2012). A prototype on a meta-model for designing instructional pervasive games. *The 4th IEEE International Conference on Digital Game and Intelligent Toy Enhanced Learning*. Symposium conducted at the meeting of Kagawa International Conference Hall and Sunport Hall Takama Takamatsu, Japan.

Chen, C. P., Guo, J. F., & Shih, J.-L. (2012). Constructing an evaluation framework for cultural-inquiry pervasive game. *The 20th International Conference on Computers in Education (ICCE 2012)*. Symposium conducted at the meeting of National Institute of Education, Singapore.

Cheok, A., Goh, K. H., Liu, W., Farbiz, F., Fong, S. W., Teo, S. L., Li, Y., & Yang, X. (2004). Human Pacman: A mobile, wide-area entertainment system based on physical, social, and ubiquitous computing. *Personal and Ubiquitous Computing*, *8*(2), 71–81.

Cisler, S. (n.d.). *Planning for sustainability: How to keep your ICT project running (schools online)*. Retrieved August 3, 2011, from http://geoinfo.uneca.org/sdiafrica/Reference/Ref6/Sustainabilit-booklet.doc

Corey, S. H. (2010). Pedagogy and place: Merging urban and environmental history with active learning. *Journal of Urban History*, *36*(1), 28–41.

Dyson, B., Griffin, L. L., & Hastie, P. (2004). Sport education, tactical games, and cooperative learning: Theoretical and pedagogical considerations. *Quest*, *56*(2), 226–240.

Edwards, R. (2003) *GNS and other matters of role-playing theory*. Retrieved on March 13, 2015 from http://www.indie-rpgs.com/articles/3/

Enyedy, N., Danish, J. A., Delacruz, G., & Kumar, M. (2012). Learning physics through play in an augmented reality environment. *Computer-Supported Collaborative Learning*, 7, 347 http://www.indie-rpgs.com/articles/3/.378.

Facer, K., Joiner, R., Stanton, D., Reid, J., Hull, R. & Kirk, D. (2004). Savannah: mobile gaming and learning? *Journal of Computer Assisted Learning*, *20*(6), 399–409.

Gentes, A., Guyot-Mbodji, A., & Demeure, I. (2010). Gaming on the move: Urban experience as a new paradigm for mobile pervasive game design. *Multimedia Systems*, *16*(1), 43–55.

Herbst, I., Braun, A. K., McCall, R., & Broll, W. (2008). TimeWarp: Interactive time travel with a mobile mixed reality game. *Proceedings of the 10th International Conference on Human Computer Interaction with Mobile Devices and Services*. Symposium conducted at the meeting of ACM, New York.

Huang, H. C., Shih, S. G., & Lai, W. C. (2011). Cooperative learning in engineering education: A game theory-based approach. *International Journal of Engineering Education*, *27*(4), 875–884.

Huizenga, J., Admiraal, W., Akkerman, S., & Dam, G. (2009). Mobile game-based learning in secondary education: Engagement, motivation and learning in a mobile city game. *Journal of Computer Assisted Learning*, *25*(4), 332–334.

Jegers, K. (2009). Pervasive gameflow –identifying and exploring the mechanisms of player enjoyment in pervasive games. Unpublished Doctoral dissertation. Symposium conducted at the meeting of Umea University, Sweden.

Johnson, L., Levine, A., Smith, R., Smythe, T., & Stone, S. (2009). *The horizon report: 2009 Australia–New Zealand edition.* Austin, TX: The New Media Consortium.

Johnson, L., Smith, R., Willis, H., Levine, A., & Haywood, K., (2011). *The 2011 horizon report.* Austin, TX: The New Media Consortium.

Kaufmann, H. (2006). The potential of augmented reality in dynamic geometry education. *The 12th International Conference on Geometry and Graphics (ISGG).* Symposium conducted at the meeting of Salvador, Brazil.

Laine, T. H., Sedano, C. A. I., Joy, M., & Sutinen, E. (2010). Critical factors for technology integration in game-based pervasive learning spaces. *IEEE Transactions on Learning Technologies, 3*(4), 294–306.

Liang, C. H. (2005). *The design and study of the across mobile platform learning system (AMPLe) supporting handheld devices in outdoor ecological learning.* Unpublished Master's thesis, National Tsing Hua University, Hsinchu, Taiwan.

Lin, F. S. (2006). *Design mobile learning instructional system about learning of campus vegetation.* Unpublished Master's thesis, National Pingtung University of Education, Pingtung, Taiwan.

Liu, Y. (2010). Social media tools as a learning resource. *Journal of Educational Technology Development and Exchange, 3*, 101–114.

Mandy, J. M. (2006). Introducing teachers and university students to GPS technology and its importance in remote sensing through GPS treasure hunts. *Journal of Geoscience Education, 54*(4), 471–475.

Manlove, S., Lazonder, A. W., & de Jong, T. (2009). Trends and issues of regulative support use during inquiry learning: Patterns from three studies. *Computers in Human Behavior, 25*(4), 795–803.

Mazman, S. G. & Usluel, Y. K. (2010). Modeling educational usage of Facebook. *Computers & Education, 55*(2), 444–453.

Mikic, F., Anido, L., Valero, E., & Picos, J. (2007). Accessibility and mobile learning standardization. *The Second International Workshop on MCL.* Symposium conducted at the meeting of Saint Luce, Martinique.

Molebash, P. (2004). Web historical inquiry projects. *Social Education, 68*(3), 226–234.

Montola, M. (2011). A ludological view on the pervasive mixed-reality game research paradigm. *Personal and Ubiquitous Computing, 15*(1), 3–12.

Niantic Labs (2013). *Ingress. [Game].* Retrieved on March 13, 2015 from https://www.ingress.com/

National Council for the Social Studies (1994). *Expectations of excellence: Curriculum standards for social studies.* Washington, DC: National Council for the Social Studies.

Ng, W., & Nicholas, H. (2013). A framework for sustainable mobile learning in schools. *British Journal of Educational Technology, 44*(5), 695–715.

O'Hear, S. (2006). *E-learning 2.0 – How Web Technologies are Shaping Education.* Retrieved on March 13, 2015 from http://www.readwriteweb.com/archives/e-learning_20.php.

Okazaki, S., Skapa, R., & Grande, I. (2008). Capturing global youth: Mobile gaming in the U.S., Spain, and the Czech Republic. *Journal of Computer-Mediated Communication, 13*(4), 827–855.

Online college (2009). *100 ways you should be using Facebook in your classroom.* Retrieved on March 13, 2015 from http://www.onlinecollege.org/2009/10/20/100-ways-you-should-be-using-Facebook-in-your-classroom/

Piekarski, W., & Thomas, B. (2002). ARQuake: the outdoor augmented reality gaming system. *Communications of the ACM*, *45*(1), 36–38.

Pragnell, M. V., Roselli, T., & Rossano, V. (2006). Can a hypermedia cooperative e-learning environment stimulate constructive collaboration? *Educational Technology & Society*, *9*(2), 119–132.

Schwabe, G., & Göth, C. (2005), Mobile learning with a mobile game: Design and motivational effects. *Journal of Computer Assisted Learning*, *21*(3), 204–216.

Shih, J. -L., & Chen, C. P. (2012). Approaching m-learning with the application of instructional pervasive game. *The 7th IEEE International Conference on Wireless, Mobile and Ubiquitous Technologies in Education* (WMUTE 2012). Symposium conducted at the meeting of Takamatsu, Japan.

Su, C. Y. (2005). Mobile learning-learning to create a new milestone. (In Chinese) *Living Technology Education*, *38*(7), 1–2.

Sweetser, P., & Wyeth, P. (2005). GameFlow: A model for evaluating player enjoyment in games. *ACM Computers in Entertainment*, *3*(3), 3–3.

Walther, B. K. (2005). Notes on the methodology of pervasive gaming. *The 4th International Conference on Entertainment Computing* (ICEC 2005). Symposium conducted at the meeting of Kwansei Gakuin University, Japan.

10 Augmented reality

Sustaining iPad-facilitated visualisation pedagogy in nursing

Hanne Wacher Kjærgaard, Lars Peter Bech Kjeldsen and Annette Rahn

Introduction

With the increase in the number of "smart" mobile devices (smart phones, iPads and other tablets) and the simultaneous development in augmented reality (AR) as a technology, a new area of mobile usage has been made possible. Previously, AR required extremely specialised and often very large equipment, but in recent years, it has become much more accessible in everyday life: any smart device has the capability of running AR apps and creating and reading simple AR "markers" through these apps. The complexity now lies more in the content that is to be shown by the AR marker.

However, up until now, the uses of AR have been restricted to a large degree to technical or technological areas – although an expansion into advertising, tourism, and museum arenas has also been seen (Hjorth, 2013). In contrast, little seems to have been done in terms of pedagogical use in school and higher education (HE) contexts (see Munnerley et al., 2012) despite the predictions of the 2011 Horizon Report, which stated:

> Augmented reality, a capability that has been around for decades, is shifting from what was once seen as a gimmick to a bonafide game-changer. The layering of information over 3D space produces a new experience of the world, sometimes referred to as 'blended reality,' (...) bringing with it (...) new opportunities for learning.
>
> (New Media Consortium & EDUCAUSE (Association), 2011)

In the following sections, the authors will first describe the background of a small case study employing iPads and AR with a clear educational/pedagogical focus. The case study will then be described, and third, problems and possibilities will be examined in terms of factors enhancing or potentially blocking the movement from project to everyday use and implementation. Readers should bear in mind that the case is a pilot project, not a top-down decided project originally intended for whole-school implementation, but rather an exploratory, experimental project. However, in the light of the positive results of the pilot study (Rahn & Kjærgaard, 2014), it seems reasonable and relevant to try to anchor and include the project in teachers' and students' everyday educational practices, and the prerequisites for this will be discussed. In conclusion, general lessons that may have been learned from the case will be summed up in order that others may benefit from the experiences.

The AR case study

Background

The project was undertaken at VIA University College, Denmark, where professional Bachelor's programs are offered in teacher education, social studies, health sciences (e.g. nursing), and engineering, and where there is a strong tradition of working on projects across faculties, especially projects relating to ICT and learning.

Against the above-mentioned backdrop of the improved accessibility of AR and the lack of studies into educational uses, VIA's research centre for ICT and learning established a small "task force" (consisting of lecturers from VIA's different faculties – ICT engineering, teacher and nursing education – as well as external technological experts) with the aim of conducting a pre-study looking into the state of the art in AR and teaching/learning, as well as into the question of which subject areas would most beneficially lend themselves to pedagogical applications of AR. The work done by the task force pointed to the affordances of AR being most immediately useful to science subjects in teacher and nursing education, for example, due to AR's ability to enhance, simulate, and clarify hidden or complex processes. As a result of this work, the project at hand was established, funded by the Faculty of Health Sciences: the use of AR as a visualising facilitator in the teaching of lung anatomy and physiology in nursing education (Rahn & Kjærgaard, 2014). The lecturer of this subject had been part of the task force.

The complexities of anatomy and physiology in nursing education

In general, the subjects of anatomy and physiology are difficult for students to fully grasp because the complexity of the human body makes it difficult to visualise the subject matter at hand and get a holistic view of the body. Through this project, one aim was to help students learn about the human body through a virtual, situated model, AR providing students with a reality as close as possible to the real human bodies of their future patients. The limited real-world context in nursing education is a general problem in higher education (HE) (Litchfield et al., 2007), but it is very pronounced in these subjects.

In the traditional anatomy and physiology classroom, student learning concerning respiration and the complex functions involved in this, e.g. lung anatomy, the exchange of oxygen and carbon dioxide, has been facilitated through models, and consequently, high levels of abstraction have been needed.

Since the experience of this area as being difficult to learn is shared by teachers, students, and places of clinical practice placement alike, there is an obvious need for change in the approaches to its teaching. Approaches and methods that facilitate student learning and provide both teachers and students with new tools and possibilities are needed.

The context

As previously mentioned, the overall context is a large university college in Denmark, and more concretely, the case unfolded at one of VIA's 6 schools of nursing. At this school, there is a generally, technology-rich environment as the school has a strategic focus on and specialises in innovation and technology (including welfare technology) in its everyday workings, both in the training of nurses at the school and in the collaboration with clinical practice at hospitals and other health institutions.

Anatomy and physiology as subjects require students to acquire knowledge of the body, including things that are invisible and inside the body and therefore hard to grasp, thus requiring a high level of abstraction: generally, students need to be able to connect a coloured plastic model, a 2D image, or a model of a skeleton to complex processes in a very live body, often also combined with a basic knowledge of physics. The pre-study indicated that one of the main affordances of AR is exactly the fact that it can render visible things that are otherwise hidden, it can help visualise complex processes, and it can also help support and underpin a modern and updated, innovative approach to the teaching of science subjects.

In terms of the case at hand, the teachers of anatomy and physiology shared the experience that many aspects are difficult for students to understand. In order to make the content more approachable to students, a pedagogy based on inquiry-based science education (IBSE) principles (Artigue et al., 2012) was chosen since it is a problem-oriented science education approach, which supports students' active, inquiring behaviour, their verbalisation of their own ideas and understandings, and furthers increased independence in their practical work. Based on social constructivist ideas (Harlen, 2013; Østergaard et al., 2010), IBSE holds that students are motivated and engaged through hands-on activities and that their construction of knowledge takes place through active thinking, i.e. selective attention, organisation of information, and integration, just as social interaction and knowledge sharing are important (Rahn & Kjærgaard, 2014). Students therefore need to be active in the teaching and learning processes and engage in deep learning as a generally more effective approach (see e.g. (Marton & Booth, 1997; Prosser & Trigwell, 1999).

In line with the IBSE approach, students in the pilot study were asked – in their work groups – to describe, validate, and discuss what they saw while using the app. Using a mobile technology in this approach is especially valuable as it allows the students to take on the needed independent and active roles in their investigation of (in this case) the lungs as they can move about, choosing the physical angle they want, choosing to see front or back, choosing to repeat or return to their "investigation" through the app whenever they feel the need (see also Litchfield et al., 2007, and Kennedy et al., 2008). In general, then, the IBSE approach is very much supported by the AR application, allowing students to explore, inquire, be active, and collaborate.

The project

In short, the project produced and tested an app showing lung anatomy *in situ*, i.e. lungs moving ("breathing") inside a body. This illusion was created through the use of the app and an AR marker on a t-shirt. When the iPad with the app was held over the marker, it was activated, and the app could thus show the front and back of lungs moving "inside" the student wearing the t-shirt.

The development phase

An external technology partner was involved, and the proof-of-concept app was produced to the specifications of the teacher and with her as quality control. Areas here that proved difficult were the procurement of reliable

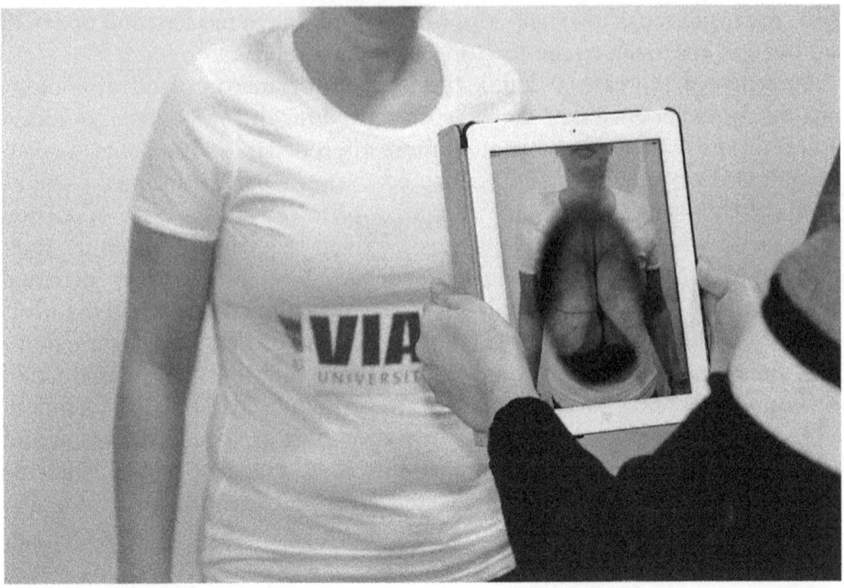

Figure 10.1 Lungs "in situ" through the iPad.

quality data, and as Högskoleverket – Swedish National Agency for Higher Education (2008) and Sjøberg(2012) say, quality is a major component in sustainability and longevity. For instance, it is essential that the materials shown/the illusion provided by the app are valid in terms of scientific knowledge concerning the body as well as the position and the functions of the lungs. Only then can the use of the app contribute[1] to valid knowledge construction with a high degree of agreement between the virtual lungs presented and the real human body; it cannot be incidental. In addition, it is essential that the app be technically stable and reliable.

The experiment

In class, students were introduced to the use of the AR app and asked to work with it on their iPads in groups in an IBSE approach, as described above. They were given iPads owned and managed by the school of nursing. Since the app was not available for download in this phase, they could not use their own devices. Thus everything was basically teacher designed and teacher led – and only existed in this particular, formal educational context.

In the survey and the interviews that followed the experiment, students were extremely positive towards the AR app and what it could add to their learning. It became clear that AR did facilitate an IBSE approach, and students reported achieving a deeper understanding of the subject matter at hand.

Among other things, they claimed that the use of AR helped them gain a more holistic view of the body. Furthermore, the visual and *in situ* characteristics of the AR used helped them "learn more", and they requested additional visual materials of this type, where they themselves could choose where to focus in relation to their individual needs. Thus new possibilities were added to their learning space, which was enhanced and allowed them more control of their own learning processes (Kjærgaard, Kjeldsen, & Asmussen, 2013).

Outcomes – and next steps?

The students involved in the pilot all expressed the opinion that they had found the use of the app helpful to their learning processes and understanding, and all judged it as more realistic than the other tools offered to them. Most saw potential in the technology as illustrated in Figures 10.2 and 10.3.

Examples of student statements showing their perceived benefit of the app are:

> [I]t has given a good picture of what it looks like when we breathe and what the lungs look like – all that with the different lobes and so on. It was also nice that you could see the back of the lungs as that is rarely seen in the books.

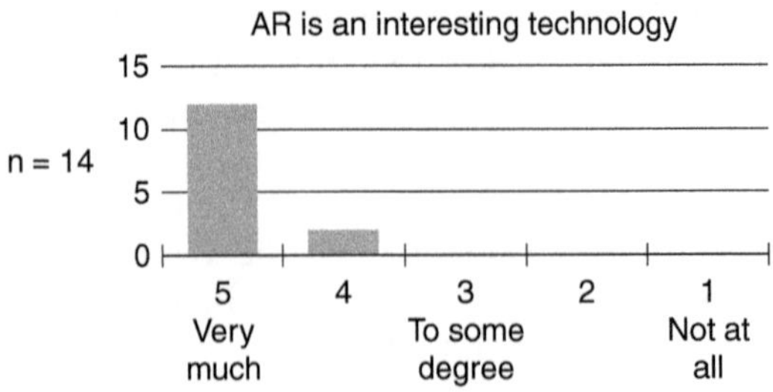

Figure 10.2 Student views of AR potential.

> [I noticed that] the lungs actually changed colour, I don't remember that from what I read. Today, I have gotten an incredibly good picture of the lungs both in inspiration and expiration.
>
> I have gotten an image of what the lung looks like inside a body and how it changes when you breathe. Better understanding of dimensions, the size and placement of the lungs, seeing front and back. [...] It has been illustrative to see the physiology and how the lungs change colour. You now have an idea of their size [...] and I now l know where they are placed and the differences in colour that take place during respiration.

From interviews, it also became clear that the use of the app, the mobile device, and the active role it allowed students to have in the learning process helped them – and they also ascribed much of the success to the visual whole that is created through the use of the t-shirt and *in situ* placement. However, students also realised the limitations of the app and were quick to suggest changes, additions and further development to help their learning:

> It is really good in the context of anatomy and physiology as this subject is very difficult to understand – it could [generally in the

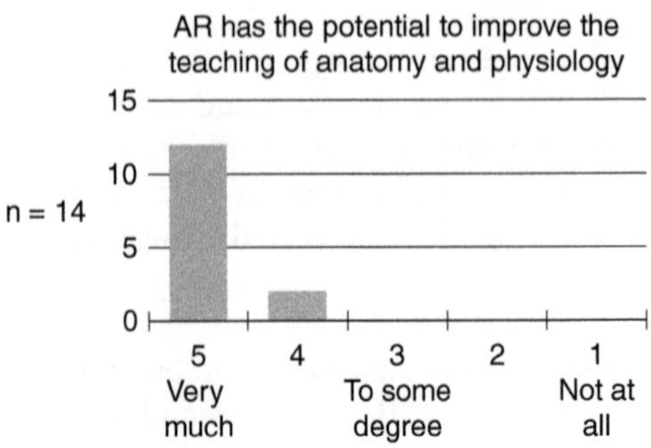

Figure 10.3 Student evaluation.

subject] create better understanding of how things work when it can be visualised in this way.

Likewise, students realised that the simple app could not adjust the size of the lungs to the person wearing the t-shirt and that it did not show real-time breathing rate – developments that would be desirable.

The project eventually came to a close when its period of funding ended, and there are challenges for its longevity and more general adoption as well as further technological development of the app. In many ways, there was initially more of a focus on experimenting than on securing sustainability.

Sustainability preconditions

There are certain preconditions that are essential and pivotal in the transition from a small successful project to a generally accepted and adopted educational practice (teaching AND learning). Cisler (n.d.) claims and describes a framework of four factors: economic, social, political, and technological sustainability. This is supplemented by Ng and Nicholas (2013) with a fifth: pedagogical sustainability.

These categories are also relevant to our context and the process and "life" of this project. Below, the project will be discussed in relation to these categories. However, in addition to the above-mentioned, a further dimension will be introduced as influential for sustainability and longevity, namely the entrepreneurial mindset of the staff involved. This will be presented under the heading entrepreneurship and briefly touched upon in relation to some of the other factors.

Economic sustainability

Unfortunately, many projects are still funded and started without them having a lasting effect on the context they arose from or related contexts, which is not economically sustainable. It might be sustainable where a project is started as experimental development or a pilot and does not live up to the initial expectations despite needs analyses, user involvement etc.

In order to even start talking about economic sustainability, it is of course necessary that the design and the product in themselves are sustainable, but not necessarily in the sense of being completely finished and immediately marketable in relation to the educational sector. These factors will be discussed later.

Concerning the economic issues, it seems clear that if the institution wants both to accommodate student wishes and support the advances and updates in how technology can enhance student learning, funding is essential in securing the first development phase. In the case studied, the technological outcome was a proof-of-concept app that was also intended as potential leverage in relation to external funding, all with the intention of reaching a technological (and pedagogical) level where the general adoption by staff and students will be made easier. Only in this way is it realistic – with a costly technological development – to sustain a project long enough for it to reach general implementation. And being willing to follow up the initial funding also needs to be in the institutional budget, both in terms of technological development and professional development of staff.

This external funding can be secured from trusts, from international project funding (e.g. the EU's research and development programs), but it can also be secured through collaboration with the professional arenas which the educational institutions collaborate with and which are the career arenas for the students after graduation. Oftentimes, advances in the application of technology in nursing education can also be used in the education of patients, giving the hospitals and other health institutions a vested interest in collaborating, also for funding.

In this section, the focus will be on technological costs as a factor that may be detrimental to a project. Additionally, professional development will be dealt with as this is essential for any change in teaching practice involving new uses of technology, and the need for user involvement will be discussed.

Technological costs

The issues in this category (Cisler, n.d.; Litchfield et al., 2007) relate to the organisation being able to overcome the costs of hardware, software, infrastructure etc. In the case described here, infrastructure as such is not a problem to be dealt with specifically, as the entire institution has a well-functioning infrastructure, and connectivity is not an issue. However, hardware purchases are a cost to be dealt with, especially in an area where the standards change fast: are iPads the right technology to choose, or are there aspects in terms of management of institutionally-owned devices that would make another (and cheaper) brand more appropriate?

Additionally, when the device is owned by the institution, technical support is a cost to be dealt with whenever there are updates to the devices, and if students' own devices are used, some cost for support to help them will also be incurred. At the moment, the app developed in this project is static and exists only on 6 devices at the school of nursing – and extra funding is necessary if it is to be useful for others and in a broader context. In fact, in order to secure that an app is freely and frequently used, it should be available to the students on their own devices and outside the classroom (Cadavieco, Goulão, & Costales, 2012).

In addition to the in-house costs, an ambitious and innovative project often requires external developer expertise. This is definitely a cost to be counted on, not only for "delivery", but also for support and backup during a testing phase. However, the cost will be most substantial in the initial phase and minimised after the developmental phase is over and a satisfactory technological solution has been reached. In the case at hand, it proved necessary to start off with a manageable proof-of-concept application in order to enable the securing of additional funding. Still, it is necessary to bear in mind that as technologies advance, add-ons or further development may be necessary to accommodate both student and teacher needs *and* the credibility of the technology. Thus, the AR project is currently at a point where additional development is necessary.

In our case, the project was funded by the Faculty of Health Sciences, but Whyte (2000) lists a number of additional funding sources, e.g. private donations, public subsidies, and one might add the use of networks and partnerships with the community (here e.g. hospitals) and businesses – thus establishing multiple sources of funding to enable the educational institution to sustain the product for an acceptable amount of time.

In order to achieve economic sustainability, it is important for a project to join forces with enterprises – private or public, including educational institutions whose main interest is not on money-making but on securing quality in education. The inclusion of these partners needs to be done at the right time – and for the project dealt with here, that might be the stage where it is currently at: there is a proof-of-concept app that might interest external partners. It may thus be possible to show the surrounding world that there is an interesting product, which is a requirement in order to get them to commit themselves to produce, host, and market the product.

Professional development

It goes without saying that if there is no investment in competence and professional development of additional staff to be involved in the spread and adoption of a new technological solution, the solution is very unlikely to catch on to any great extent in the organisation. As new technologies or new uses of old technologies are often contingent on a change in pedagogical practice and maybe also a change of pedagogical approach and of the understanding of the roles of teachers and students, general adoption requires both competence development and alignment of pedagogical aims and goals in order to provide quality adoption. In the case dealt with here, the use of AR is very much attuned to a distinct pedagogical approach, namely that of IBSE, which again requires teachers to agree with the advantages embodied in this pedagogy, for example active students, inquisitive students, etc.

This professional development (PD) does not come about naturally – it needs both organisational and financial assistance. The organisation can do a lot to ensure that teachers become ready to take on the adoption of new technologies, e.g. through strategic goals, a clearly communicated focus, and investment in technology. Teachers' beliefs, readiness, and the availability of technology are direct, positive influences on technology integration (Inan & Lowther, 2010). In order to ensure the success of PD, the organisation needs to assess the needs of individual staff members so that differences in their previous experiences are recognised (Kjærgaard & Kjeldsen, 2011). Thus, a differentiated model needs to be adopted. Additionally, the organisation needs to make room for and support collaboration between teachers in connection with PD (Kjeldsen, 2003; Prinds, 1999).

As it is a work in progress, the case described is not yet at a stage where the app is ready for broader implementation. However, if the project is to become successful, PD needs to be dealt with before long, for instance among the 6 schools of nursing at VIA or at national level. This may be done in terms of either workshops or short courses provided by the school for its own employees or at national level by the School of Continuing Education. As the use of the app is particularly supportive of student-activating pedagogical approaches, it is not just a question of teaching staff using the app; the whole approach to teaching that goes with it also needs to be part of the agenda.

User involvement

It is self-evident that the development of technological products or solutions that have no demand will never achieve funding. Often, when applying for funding, whether in one's own organisation or from other sources, there is a strict requirement for a needs analysis and a description of potential users. Despite this, many projects fail to give priority to establishing a relationship with the users just as these are not necessarily involved in the development processes. In order to achieve a high degree of sustainability, it seems essential that users are involved and can see their own potential gains from the project. This may be done in various ways, but in the development of a service or product to be used in an educational context, students and colleagues who are not part of the core project group need to be involved, e.g. through hands-on sessions, workshops during the development phase (for input) or internal conference presentations of project and results.

Until now, the project has not involved users until the phase where the app was presented. In order for the project to be successful in its further development, it is now at a stage where student and lecturer voices need to be heard – and "critical friends" from outside the primary context will also be able to provide valuable input for the further development of the app.

'Also, user involvement may be said to be particularly important in cases where justification and support is needed for further funding, e.g. for further development, expansion, or design development together with external (business) partners.

Social Sustainability

Cisler (n.d.) describes social sustainability as related to the wider community: "[l]istening to the people most affected by the change". Cisler is here in line with McKenzie (2004), who defines social sustainability as: "a life-enhancing condition within communities, and a process within communities that can achieve that condition".

In this case, the people affected by the change would include students, nursing teachers, and partners of various kinds that may all benefit from the end results, e.g. hospitals and patients. In addition to these "user partnerships", also "developer partnerships" are necessary.

User partnerships

In order to be in a partnership, users need to buy in to the project (Ali & Bailur, 2007). This is a two-way process: the project needs to approach relevant users, who, in turn, need to commit themselves. The project doesn't have to show its general relevance to different user groups but should perhaps rather focus on a few relevant users (Heeks, 2005). The user partnerships that are relevant for sustainability may be internal or external: in the case at hand, internal users whom it would be desirable to involve and commit would be other nursing teachers and, not least, the students. In order to ensure the involvement of these groups, it is of course necessary to be able to communicate to them both the affordances of the app and the presumed match with student needs: students will be accommodated here by the app being mobile, allowing them flexible and thus more equal access, just as it allows for student learning approaches that are more visual and interactive – when this is what individual students need. Likewise, teachers are given the opportunity to accommodate these needs through the app.

External partners might be hospitals, as these will benefit from student nurses with a deeper understanding of anatomical and physiological complexes, but also with new knowledge, not least concerning how to communicate with patients in patient education. As an example, the project was presented to the staff at the places of clinical practice for all of VIA's schools of nursing with great interest as a consequence. This shows the importance of the involvement of user partners.

Developer partnerships

In the project at hand, there was an internal partnership for development that involved a research unit, an educational unit (school of nursing), a faculty, and moreover there was an external partner: a GTS institute[2] with the appropriate technological knowhow. This partner was, however, funded by the internal partnership.

In projects with several partners, the strength of the partnership may be measured in two ways. First, the strength relies on the proper competences being present; second, strong commitment of all partners is necessary. In the AR project, the first version was in some ways merely a "proof-of-concept" development that would allow the entire consortium to apply for external funding for a larger project. However, since the results in the piloting phase were very positive, it became relevant to implement and spread the existing prototype in a slightly updated version, which, in turn, made demands on the commitment of the partners, ensuring a stable version that can be made accessible from various app stores.

In the project, one way of achieving sustainability could be that one of the developer partners made available their technological systems, know-how, and platforms. This, again, would require a high degree of organisational commitment, alignment of commitments, and a willingness of this partner to be visionary in terms of sustaining the project results and technological solution – both during and after the project phase, and also in terms of marketing the solution or product developed.

An additional element that is important for social sustainability is the reputation of the product and its provenance: is the product related to an organisation that is perceived as having clear ethical values, being trustworthy in terms of "brand" and providing value for money? In the pedagogical context of this project, these factors might be met by the organisation's mission to create value and better health conditions for patients, the product's pedagogical approach, design and reliability.

Political sustainability

This issue will be discussed as consisting of organisational support, which is important as sustainable projects cannot be the projects of individual "free agents"; of visions, i.e. as supporting strategies and sense-making at various levels; and of an organisational focus on knowledge sharing, essential for anchoring the project in its immediate environ and providing it with visibility for potential stakeholders.

In general, the organisation needs to be able to formulate statements like: we embark on this project because we expect it to have [a specific] value for us and because it supports [a specific] strategy and activity in the organisation. In other words, the organisation as a whole needs to assume ownership of the project and the product. And once a project is given support, the organisation also needs to focus on keeping momentum (Ng & Nicholas, 2013).

Organisational support

In this case, three organisational units were influential in providing opportunities for sustainability: Research Centre for ICT and Learning, Faculty of Health Sciences, and the School of Nursing.

To begin with, the support of the Research Centre for ICT and Learning lies in the establishment of the pre-study through funding and organisational setup. This appears to have been a strong driver at the outset. Thus the centre facilitated the setup of the task force, providing financial and research support. During the project phase, the centre provided research support and support for article writing and conference participation.

In the project phase, the Faculty of Health Sciences has provided organisational support in the form of staff and funding, just as a framework for internal knowledge sharing has been provided through internal conferences, support for conference participation etc.

Support from the School of Nursing was provided in the form of supporting teacher interests and visions, and in the form of sustaining the teacher's arguments for the potential benefits for the students and the necessity of the solution. This, in turn, meant that the teacher was given room to pursue her interest and present results in her local environment.

Visions

Organisational visions are important for the sustainability of a project: if the organisation does not see the project as supporting these and assumes ownership or responsibility, little will be gained or achieved.

The Research Centre for ICT and Learning had initial visions of promoting research into the possibilities and uses of AR in education. Therefore, there is a natural interest in securing external funding for a period of time sufficient for the development of real change in and innovation of teaching practices. In terms of securing the longevity of the solution, a research interest project may be seen as beneficial: a research facility will have an inherent interest in following the project, aiming for a return on its initial investment, both in terms of money and research results and direction.

The overall vision of the Faculty of Health Sciences is to underpin the development of and change in educational content and methodology, aiming at a modernised education that will allow graduates to contribute to development and necessary change in the professional field. There is an additional strong interest in supporting projects that can help strengthen the connections between research, education, and profession, adding value to all three.

In a pedagogically aimed project, it is essential to have a visionary teacher as a driving force – or at least as a willing participant in an organisationally prioritised project.

In our case pedagogical visions of the teacher from the School of Nursing were a strong driver for change and development in her subject area through the work with the development and test of the app as part of a new pedagogical approach. She kept a clear focus on the affordances of the technology as a helper in facilitating and enhancing student learning as well as of its appropriateness for an updated approach to science education (IBSE and active, collaborative learning).

Knowledge sharing

A project that develops new ways of teaching a certain professional content and develops new general possibilities for and change in pedagogical methods needs to share its knowledge as one component in allowing – and persuading – others to use it. Here, the positive outcomes and the perceived advantages of students gave the project a strong platform for knowledge sharing although users (other teachers or students) had not been much involved earlier in the project.

The organisation can do much to promote the results of a project and secure an interest in other parts of the organisation, simply by providing organisational space for its presentation and time for the people involved to actually spread the results. In our case, this has been done through internal faculty conferences and presentations at other relevant schools of nursing. However, if the project is to have longevity, it also needs to be allowed external exposure: the project at hand has resulted in articles and conference presentations both nationally and internationally, supported by the organisations involved. An external interest – whether from partners, the wider community, or the international research community – will naturally add to the likelihood of the organisation investing further resources in securing the longevity of the project and in its ability to attract further funding for additional development that may increase its quality and applicability. This, in turn, may encourage additional partners to take on the project and product as a business.

To increase the impact of a project, the dissemination and discussion of results, drawing national or even international attention to it, using *Facebook* and *Twitter* as channels of dissemination and "marketing" is also a possibility. Additionally, these channels of communication are free for all partners in a project to use, thus ensuring that the project and its results do not end up staying within the walls of the institution.

Technological sustainability

As mentioned earlier, the basic cost and complexity of AR as a technology has been substantially reduced in recent years. That, in itself, means that the technology is more approachable and immediately useful for a broader section of the population. In general, students at this college are required to bring their own devices, but as previously mentioned, that was not possible with the *iPads* in the pilot case. However, if the AR apps are to be sustainable, there is no doubt that the students need to have ease of access and be able to both use their own devices *and* use the app with flexibility of time and place according to their study needs. This also means that the product needs to be available on platforms that are accessible to the students and part of their "normal" behaviour (Cadavieco et al., 2012) (see also earlier in this chapter).

Still, because the anatomical/physiological content is quite demanding in the "production phase", funding is needed for an external, technological partner. This funding needs to be available not once, but several times to ensure development and to make sure that the product develops from a mere limited prototype with only one anatomical focus to a broader product involving more areas of the anatomical curriculum as needed by the students and lecturers. In order to secure this, both organisational and economic sustainability are involved: this requires a potential partnership with an external agent that can make a business case based on the preliminary results. In our case, this might be a publisher or a partner who can provide the technological development needed *and* market and sell the end products in an educational context, whether in the training of health professionals or of patients and their relatives.

Pedagogical sustainability

A basic prerequisite for any success in an educational context is that the staff involved in the development of an innovation or R&D project are highly qualified in the areas that the project addresses. This must be recognised by the users (here the students, additional teaching staff, and potentially hospitals and patients) in order for them to accept the technological solution as valid and trustworthy. Having what one believes to be a good solution, product or method is of little use if the driving and implementing force of the project is not recognised as having a capacity in his or her field – here, as having a capacity within content, pedagogical methodology, and technological knowledge. This is true for all partners who contribute to a project, but the person in the front line is the one that will be the target of doubts and questions. Therefore, it is very important that this person knows and can mediate the exact strengths and advantages of using this particular product/method over another, especially if the product/method is innovative and breaks with previous practices and expectations of more traditional teaching.

Furthermore, to achieve pedagogical sustainability, it is necessary to involve the users, in this case the students, in a longer perspective, perhaps also the clinical placement stations. In the case described here, students were invited to provide their view of future needs, both in terms of content and functionality, providing new input for the further technological development that is necessary for the project to live on and be spread to other areas of the organisation. Also, when the product is spread to other lecturers, their ideas will need to be recognised and, if possible, given life in a further development.

In order to achieve pedagogical sustainability, it is also important that the managerial level supports the strategically "right" projects. This presupposes a reasonable level of knowledge about pedagogical trends, technologies and the potential pedagogical gains achievable through these. In this sense, the project and product at hand supports the trend towards the creation of learning objects that students may use in informal contexts, reflecting the increasing focus on informal learning arenas as seen in, for example, the MOOC movements, in flipped-classroom approaches and their connections to the formal arenas.

Also the teacher is important in several ways: in terms of the development of the technological solution, the teacher is responsible for quality assurance as the technologists are completely dependent on this; they can only deliver the technology, not ensure the subject-based and pedagogical quality. In our instance, the teacher was essential in requesting necessary changes in, for example, the colouring, size, and position of the lungs, ensuring that the virtual lungs appeared in a correct and natural manner. In our project, where the main goal was to facilitate student learning, this was much more important than the use of colours and movements that might have "sold" the technology better in an open market.

As teachers are central to student achievement (Hattie, 2009), it is of course also essential that the teachers whom the organisation would like to take on the product and method developed are provided with professional development – i.e. involving the factors pertaining to organisational and economic sustainability.

Entrepreneurship

Sustainability has been a focus for many years as no one can condone the continued start – and death – of numerous projects that never find a sustainable life. Although all the previously mentioned factors are essential for sustainability of ICT projects, there still may be a need to re-think the role of the teacher.

Traditionally, the teacher's role in securing project longevity has perhaps been underestimated: this has been seen as the role of the organisation rather than the individual. But as has been the experience in this project and others like it, there may be a need for the teacher to take on a more entrepreneurial role.

As part of one's professional life as a teacher, the opportunity often arises that allows one to establish good relations with businesses, developers, or distributors of technological solutions. Or maybe a colleague has an extended network that comprises relevant actors in the field. In fact, teachers should not abstain from taking advantage of these contacts. If you believe in your product or solution, and if you have the support of your organisation, you also have a good case in relation these other actors. However, it is not natural for a teacher to consider the results of his efforts as merchandise, but in order to achieve success in relation to spreading one's products outside of one's own classroom, maybe even to other target groups or groups of users, it is necessary to be willing to learn to do this. After all, many of the teachers in question are innovators or early adopters (Rogers, 2003) that need the support of their organisation in order to be able to help drive change on a larger scale. It is essential for the teacher involved to be able to present (or market) the qualities of the product developed, to be proactive and persistent.

Teachers should not underestimate the value of boosting their name through the credits they have earned in the course of their professional career. This is true of academic as well as experience-based results.

Conclusion

Looking at the project through the lens of the categories described above has allowed the authors to see perspectives for the future life of the project. Some of the foundations for sustainability and longevity may have been laid, but there are also areas where we need to act now in order to secure the future life of our project and product. Among these are: an increased focus on user involvement, addressing the issue of a business partnership, and an even stronger focus on communication and knowledge sharing in relation to the project.

However, we also see potential areas that we think need to be included in dealing with sustainability preconditions, namely the need for teachers to take on a new view of their own role as project makers. Teachers need to act in ways that may not be part of their immediate understanding of their profession and daily tasks. Teachers are not used to thinking in terms of marketing and "sale" of ideas and products – they deal in knowledge rather than money.

Maybe there is a dilemma in this: teachers *do* need to maintain a strong focus on providing quality teaching, but in order to render possible the sustainability and success of mobile ICT development projects meant to help provide this quality, they also need to have the entrepreneurial mindset described above. But maybe it is asking too much that teachers need to play a central role in both project development and implementation *and* in the securing of sustainability?

Notes

1 It should be noted that students also had access to other materials about lung anatomy and physiology; the app was a supplementary tool.

2 See http://www.teknologiportalen.dk/EN/About+GTS/About+the+GTS+institutes/ for a description of these technology and innovation centers.

References

Ali, M. and Bailur, S. (2007). The challenge of "sustainability" in ICT4D – Is *bricolage* the answer? *Proceedings of the 9th International Conference on Social Implications of Computers in Developing Countries*, São Paulo, Brazil.

Artigue, M., Dillon, J., Harlen, W., & Léna, P. (2012). Learning through inquiry. *The Fibonacci Project.* Retrieved on December 7, 2014 from www.fibonacci-project.eu

Cadavieco, J. F., Goulão, M. de, F., & Costales, A. F. (2012). Using augmented reality and m-learning to optimize students' performance in higher education.

Procedia – Social and Behavioral Sciences, *46*, 2970–2977. doi:10.1016/j.sbspro.2012.05.599

Cisler, S. (n.d.). Planning for sustainability: How to keep your ICT project running (schools online). Retrieved November 3, 2011, from http://geoinfo.uneca.org/sdiafrica/Reference/Ref6/Sustainabilit-booklet.doc

Harlen, W. (2013). *Assessment & inquiry-based science education: Issues in policy and practice*. Trieste: Global Network of Science Academies.

Hattie, J. A. C. (2009). *Visible learning: A synthesis of over 800 meta-analyses relating to achievement*. London: Routledge.

Heeks, R. (2005). Sustainability and the future of e-development. *eDevelopment Briefing* No. 10. Institute for Development Policy and Management, University of Manchester. Retrieved on December 7, 2014 from http://www.sed.manchester.ac.uk/idpm/publications/wp/di/short/DIGBriefing10Sustain.doc

Hjorth, M. (2013). *Augmented reality – pedagogical opportunities*. Aarhus: VIA University College, Research Centre for E-Learning and Media.

Högskoleverket – Swedish National Agency for Higher Education. (2008). *E-learning quality aspects and criteria for evaluation of e-learning in higher education*. Högskoleverket – Swedish National Agency for Higher Education. Retrieved on December 7, 2014 from http://www.hsv.se/download/18.8f0e4c9119e2b4a60c800028057/0811R.pdf

Inan, F. A., & Lowther, D. L. (2010). Factors affecting technology integration in K-12 classrooms: A path model. *Educational Technology Research and Development*, *58*(2), 137–154.

Kennedy, G. E., Judd, T. S., Churchward, A., Gray, K., & Krause, K.-L. (2008). First year students' experiences with technology: Are they really digital natives? *Australasian journal of educational technology*, *24*(1), 108–122.

Kjærgaard, H. W., & Kjeldsen, L. P. B. (2011). A context-based approach to e-learning quality. *E-learning innovative models for the integration of education, technology and research. 5th GUIDE International Conference Proceedings*. Rome: Guide Association.

Kjærgaard, H. W., Kjeldsen, L. P. B., & Asmussen, J. B. (2013). The convergent learning space: An updated learning space for 21st-century teaching and learning. *I World Conference on E-Learning in Corporate, Government, Healthcare, and Higher Education*. (Bd. 2013, s. 1971–1976).

Kjeldsen, L. P. B. (2003). *Når skolen udvikler sig indefra*. Copenhagen: Danmarks Pædagogiske Universitetsforlag.

Litchfield, A., Dyson, L., Lawrence, E., & Bachfischer, E. (2007). Directions for mlearning research to enhance active learning. *ascilite 2007: ICT: Providing Choices for Learners and Learning*, Singapore: Nanyang Technological University.

Marton, F., & Booth, S. A. (1997). *Learning and awareness*. Hove: Psychology Press.

McKenzie, Stephen. (2004). *Social sustainability: Towards some definitions*. Hawke Research Institute, University of South Australia.

Munnerley, D., Bacon, M., Wilson, A., Steele, J., Hedberg, J., & Fitzgerald, R. (2012). Confronting an augmented reality. *Research in Learning Technology*, *20*.

New Media Consortium, & EDUCAUSE (Association). (2011). *The Horizon Report 2011*. Austin, TX; Boulder, CO: The New Media Consortium; EDUCAUSE Learning Initiative. Retrieved on December 7, 2014 from http://www.nmc.org/pdf/2011-Horizon-Report.pdf

Ng, W., & Nicholas, H. (2013). A framework for sustainable mobile learning in schools. *British Journal of Educational Technology*, 44(5), 695–715. doi:10.1111/j.1467-8535.2012.01359.x

Østergaard, L.D., Sillasen, M., Hagelskær, J., & Bavnhøj, H. (2010). Inquiry-based science education – har naturfagsundervisningen i Danmark brug for det? *MONA-Matematik-og Naturfagsdidaktik*, *4*.

Prinds, E. (1999). *Rum til læring*. Ringsted: Centre for Teknologistøttet uddannelse.

Prosser, M., & Trigwell, K. (1999). *Understanding learning and teaching: The experience in higher education*. Buckingham: Open University Press.

Rahn, A., & Kjærgaard, H. W. (2014). Augmented reality as a visualizing facilitator in nursing education. *INTED2014 Proceedings* (s. 6560–6568). IATED.

Rogers, E. M. (2003). *Diffusion of innovations* (5th ed.). New York: Free Press.

Sjøberg, S. (2012). *Naturfag som almendannelse: En kritisk fagdidaktik* (2. udg.). Århus: Klim.

Whyte, A. V. (2000). *Assessing community telecentres: Guidelines for researchers*. Ottawa: International Development Research Centre.

11 Achieving sustainable mobile learning through student-owned devices and student-generated multimedia content

Laurel Evelyn Dyson

Introduction

Despite a growing interest in mobile learning by both teachers and researchers, there yet remain many issues to be resolved before it becomes embedded in the normal educational practice of most institutions. As Ng and Nicholas (2013, p. 696) note, few mobile learning studies "conceptualise sustainable learning", and indeed few have sustainability as a central aim. Rossiter and Crock (2006, p. 287) remind us that sustainability has at its core "the capacity for an innovation ... to grow and mature (rather than merely survive) within a particular system, without undue reliance on external interventions". It thus embodies the concept of long-term viability, rather than the typically limited duration of a subsidised project.

Achieving sustainability and knowing when it has been achieved are, no doubt, complex issues. Sustainability is not a one-dimensional concept but incorporates notions of widespread usage, integration into the institutional culture, legitimisation, and evaluation as to its effectiveness (Rossiter & Crock, 2006). However, evaluation of sustainability is often difficult. Traxler and Vosloo (2014) write of the paradox that evidence for sustainability is most often claimed from small-scale projects with short-term funding, which can tell us nothing of sustainability over time or of the potential of mobile learning on a larger scale. They further critique how mobile learning is usually evaluated in schools rather than outside school; and in formal educational settings rather than in informal or lifelong learning contexts. The greater difficulty of rigorous evaluation outside the controlled environment of the institutional setting is obviously the reason for this bias.

The literature cites a number of barriers which contribute to the lack of sustainability of mobile learning. Certainly costs associated with implementing mobile learning are often cited, and these include the more defined expense of purchasing mobile devices and off-the-shelf applications, as well as usage charges which, due to their unpredictability, educational institutions are rarely willing to fund (Dyson et al., 2009). The rapid evolution of mobile technology exacerbates this issue, with devices becoming obsolete in a short time span, so that choice of and support for technology becomes an issue (Traxler, 2005). In addition, teachers often have little money to pay for the development of content or applications, and seldom have the time or skills to create the sophisticated multimedia resources that the current generation of students increasingly expects. Whilst there is a wide range of multimedia resources and apps available for free or at only a small cost for school education, the availability of mobile learning resources suitable for higher education is less clear. Given the more specialised nature of university courses and the tendency of academics not to follow a standardised curriculum, but instead to build their courses around their own research and expertise, there is a need to develop new multimedia content for university-level programs. Thus there is a pedagogical issue, both from the teachers' point of view but also with regard to students' preferred ways of learning. These barriers fall roughly into three aspects of Ng and Nicholas' (2013) framework for sustainable mobile learning: economic sustainability, technological sustainability and pedagogical sustainability. These will be the focus of this chapter.

Notwithstanding the previously mentioned obstacles to sustainable practice, in the sphere of higher education these barriers can be overcome. On the one hand, almost all university students own a mobile device, such as a smartphone or tablet computer, which owing to convergence possesses a range of functions which lend themselves to various learning activities. This can reduce the need for institutions to buy devices in the first place or replace devices when they become obsolete (Dyson et al., 2009). On the other hand, the interest and capabilities of our students in creating user-generated content in their private lives (for example, uploading photos and videos to social networking sites such as *Facebook* and *YouTube*) can be harnessed by having students create podcasts and other content as part of rich learning activities and also for later re-use as resources by their peers (Dyson, 2012). This can save the time and cost of producing or buying multimedia resources while keying into the power of peer learning. It further reduces the need for teachers to develop the multimedia production skills required to make resources which provide a compelling listening or viewing experience.

To illustrate these opportunities, two examples from the author's research and teaching will be given. Both of these have resulted in sustainable practice over a period of several years since they were initially introduced, and well beyond the period of initial project funding. The first case study examines the Bring Your Own Device (BYOD) strategy where accounting students use their own smartphones, laptops or tablets as an alternative to "clickers" to respond to questions in lectures. In this way students have become more actively engaged in their learning and gained formative feedback. The second case study involves student-generated multimedia content where Information Technology (IT) students create vodcasts (video podcasts) of interviews with IT professionals as part of an assignment to enable students to learn more about careers in their future profession and to contribute to an ongoing resource library of career vodcasts. These case studies provide two examples of how to move forward in seeking sustainable mobile learning in the university sector.

Opportunities for sustainable mobile learning

Bring Your Own Device (BYOD)

BYOD has gained a growing number of advocates over recent years. Prensky (2004) was one of the first to suggest using students' own mobile phones in education, posing the challenge: "How to use the 1.5 billion computers already in our students' and trainees' pockets to increase learning?" The technological development of easy-to-use, multifunction, networked mobile devices at low cost, together with practically ubiquitous WiFi and cellular networks, has resulted in their widespread adoption by consumers and their embedding in personal and business use. Now that nearly all university students own a mobile device of some kind, there is a huge opportunity to extend this practice to the educational sphere.

The shift from university supplied devices to BYOD has several advantages. Most obviously it saves on the purchase price of equipment for student use, but student-owned devices inevitably come with software, storage and access to a telecommunications provider, so that the savings extend beyond the actual hardware. One university IT administrator noted, it is "really BYOE ('E' for everything)" (McPherson, quoted in German, 2013, p. 24). Moreover, the devices that students are bringing with them to university replicate to a large extent the capabilities of the technology provided by universities, technology which is largely paid for by students themselves in their tuition fees. Given this duplication, the university's provision of desktop computers does not represent a sensible allocation of university funds. Instead, German (2013, p. 18) recommends that, "[w]e need to embrace BYOD not to save money but to be able to spend money instead on specific technical capabilities that our community members really need from us and that they find unquestionably valuable".

The BYOD movement represents a change in the way that universities think about students and technology. It replaces the standardised roll-out of desktop computing and fixed-line connections with a more personalised approach which reflects students' preferences in terms of device type, platform, apps (applications), interaction styles and ubiquitous connectivity. Campbell (quoted in German, 2013, p. 12) sees it as promoting "agency, customization, and improvisation" and an expression of the "meaningful learner individuation" that we associate with a democratic system. As such, BYOD, as a means of addressing concerns about the cost of mobile learning, also addresses the needs of students living in a mobile world. At this time, it means mobile learning with smartphones, since this is the number one device in terms of student ownership (Norris, & Soloway, 2011), but in the future this could well change since the sale of tablet PCs and iPads is growing rapidly, with these devices being now owned by more than half of Australian adults, for example (AIMIA, 2013).

Shifting to a BYOD strategy is not without its potential difficulties, of which the various stakeholders will need to be aware. Some challenges highlighted in the research literature include (Adams, 2012; German, 2013; Patten & Harris, 2013):

- the need for greater bandwidth as students potentially connect to the university wireless network with multiple devices simultaneously;
- issues of providing support for a diversity of devices and operating systems;
- how to maintain security of data when the devices accessing the network are outside the control of the IT administration and often poorly managed by students from a security point of view; and
- the equity issue, if students from low socio-economic backgrounds do not own devices required for the mobile learning activities designed by their teacher.

Some of these issues will be addressed in the BYOD case study presented later.

Student-generated multimedia content

There is an increasing interest in student-generated multimedia content, paralleling the rise in user-generated content on the Web. Technological convergence has, no doubt, been a major enabler of this phenomenon (Dyson, 2012). Smartphones and tablet PCs or iPads are prime examples of the converged device, combining computing, communication, Internet connectivity and recording functions. These characteristics allow for:

- capture of content in and out of the classroom;
- short-term storage of this content on the device;
- sharing of content with other students and with the teacher; and
- the ability of most students to purchase a device in the first place (Dyson, 2012).

These devices have an affordance "as tools for complex and sustained tasks and problem solving", including authentic learning activities, data gathering in the field and the creation of multimedia content (Herrington & Herrington, 2007, p. 7).

Many approaches to student-generated multimedia have been developed, although some have been more widely adopted than others. One of the earliest and most successful approaches was digital storytelling (Lambert, 2002), a very powerful way of engaging students, more often used in school education than higher education. Slowmations are a more recent approach, a type of simple stop-go animation, which has been developed for use in primary school science education or for educating trainee primary school teachers about science concepts (Hoban, Loughran, & Nielsen, 2011; Hoban & Nielsen, 2013, 2014). Screencasts are a form of multimedia content which began as expert-generated but have started to shift in recent years to student-generated (Mohorovičić, 2012; Rocha, & Coutinho, 2011; Shafer, 2010). Podcasts and vodcasts are likewise media most often made by experts, either teachers or content developers, for students to listen to or view. More recently, examples of student-generated pod- and vodcasts have appeared in the literature (Dyson, 2014; Lee, McLoughlin, & Chan, 2008; McGarr, 2009).

The contribution of student-generated multimedia content to sustainability may be examined from two different perspectives. First, it reduces the cost to the educational institution of buying or commissioning multimedia content for the creation of learning resources and thus makes mobile learning more sustainable economically. Second, given that institutions seldom have sufficient money to pay for the development of custom-made multimedia resources and the burden for this more often than not falls on the teacher, it contributes to pedagogical sustainability in saving teacher time: both the time for teachers to develop the skills in multimedia design, production and editing, as well as the actual time producing the resources once the teacher possesses the necessary skills.

Like the BYOD approach, student-generated multimedia projects recognise that students' everyday technology preferences and practices should be incorporated into their learning. Students who have been exposed to technology for most of their lives require new pedagogical methods to engage them (Tapscott, 1998). Students' interest in and skills for producing user-generated multimedia in their spare time can be harnessed in creating student-generated learning resources for teaching discipline-specific content to their peers.

In order to do this well, they must have a deep understanding of the topic since to explain a concept one must know it (Frydenberg, 2006; McGarr 2009). Some authors go so far as to suggest that it is the very awareness of their peer audience that spurs students on to produce their best work in order to maintain the respect of their fellow students (Wheeler, Yeomans, & Wheeler, 2008). The authentic voice of the students emerges as they express their creativity, sense of humour and individuality (Frydenberg, 2006).

Case study 1: A BYOD strategy to replace clickers in large lectures

Background: Interactive classroom systems

Interactive classroom systems have been in use at universities since the 1960s for improving student engagement, interactivity and formative feedback in large lectures (Judson & Sawada, 2002). They are known under a number of names, including personal response systems, electronic voting systems or simply "clickers", from the wireless devices that students regularly use to input their answers. They allow the instructor to put a question to the class which students answer electronically, and the combined class response is displayed on a large screen in the lecture hall, usually in the form of a histogram. This is followed by immediate feedback from the instructor and if necessary questions from the students and a discussion. These interactive systems recast an otherwise usually passive, teacher-centred instructional situation (where the lecturer talks and students listen) into a more active learning process. They have been deployed in a range of disciplines where large lectures are the norm, such as business, science, medicine, information technology and engineering (Carnaghan et al., 2011; Draper, & Brown, 2004; Laxman, 2011).

Despite their many pedagogical benefits, their use in large lectures is still comparatively low (Carnaghan et al., 2011). There are several issues with the current clicker-based technology which act as major disincentives to academics. Firstly, cost is an issue, both in terms of the price of the clickers and receivers as well as the cost of technical support to install and maintain the software and hardware (Mundy, Stephens, & Dykes, 2010). If the university is subscribing to a commercial provider, then the cost of subscription (which includes access to the management system for uploading questions and viewing individual student performance, as well as technical support from the vendor) needs to be taken into account. Moreover, academics may find the management software complex to use, particularly as vendors introduce more functionality into their systems: Carnaghan et al. (2011, p. 279) note that "[w]hile the range of options allows the instructor great latitude in how to personalize usage, learning how to access and effectively use all the possibilities can be challenging". Perhaps the most important challenge associated with clickers is their cumbersome nature: carrying enough clickers to class to serve all the students and the time taken to distribute them at the beginning and collect them at the end of the lecture may be beyond the capabilities of the lecturer unless a teaching assistant is employed, which obviously adds to the cost (Freeman et al., 2007).

A more sustainable approach to interactive lectures is asking students to respond using their own technology. This eliminates the cost of buying clickers, the logistics and security challenges with the issue and return of devices, and the technical problems reported with software installation (Freeman et al., 2007). One approach is to collect students' responses to short answer or multiple-choice questions via SMS, with students texting from their mobile phones to a system installed on the lecturer's computer (Lindquist et al. 2007; Scornavacca, Huff, & Marshall, 2009). An alternative approach, that has gained traction in recent years with the rise in student ownership of smartphones, is for students to communicate via the Internet. The students usually log onto a designated website to answer questions using their smartphone, tablet or laptop, or less commonly they use a smartphone app to answer (Rubner, 2012). If lecture halls have access to a campus-wide wireless network, students can answer questions for free, or otherwise pay a small charge to their mobile service provider. Lecturers also access the system via a website to upload questions prior to class, and in the lecture activate the display of class results directly from their slides using an embedded link. Some commercial vendors of clicker-based systems are now offering proprietary Internet-based systems as an alternative (Carnaghan et al., 2011). In addition, there are several Internet-based systems reported in the literature that have been developed by universities or research institutions, some of them restricted to in-house use (e.g., *mbclick*: Rubner, 2012), while others are available to anyone for free (*VotaPedia*: Habel, 2011).

The case: An Internet-based interactive classroom system using students' own devices

The case reported here concerns an Internet-based interactive system that was first developed in late 2007 at the author's university. In 2008 it was implemented in an introductory accounting course and has been used continuously ever since. Because the subject in which it was introduced was very large, with one two-hour lecture per week repeated across four groups of approximately 300–400 students each, the lecturer was concerned about the need for a system that was simple to use, both from his and the students' point of view. He recognised that, with such large numbers of students taking the subject, the logistics of distributing and collecting clickers before and after the lecture would be too unwieldy.

In 2008 most students owned mobile phones, so that asking them to use their phones or laptops to answer questions seemed a reasonable alternative. A major Australian survey of first-year university students at the time had shown that 97.3% owned or had access to a mobile phone (Kennedy et al., 2008). Despite this, the lecturer was disappointed the first semester the system was introduced to find that on average only about 15% of the class used the system to answer questions, with a maximum of 33% electronic participation at any one time. Various reasons were given by students when surveyed as to why they were not logging in: some did not have Internet-enabled mobile phones or laptops; some were put off by concerns over cost since they believed that all Internet use on mobile phones was very expensive; while others did not know how to log onto the Internet from their phones as they had never done it before or were not even aware as to whether their mobiles were Internet-enabled or not (Dyson et al., 2009). However, there was a high level of satisfaction with the system, since students who did not answer electronically often participated by answering "in their head" and appreciated the feedback and interactivity. So, in spite of the low electronic participation, the BYOD strategy was deemed to have been sufficiently successful to warrant its continuation.

In subsequent years, student ownership of mobile devices with Internet capability has increased and students have begun to access the Internet from their devices as a matter of everyday practice. A survey of the first-year accounting students in 2013 showed that students brought the following mobile devices to lectures: 99.9% brought a mobile phone, 45.6% brought a laptop, and 17.2% brought a tablet PC or iPad. Only 3.7% of the students had no Internet-enabled device with them in lectures. When asked how often they normally accessed the Internet from their phone, most stated that they accessed it several times every day. As might logically be expected from these new figures, the level of electronic participation in class with the interactive system grew enormously compared to 2008. At the end of semester, only 6.5% of students stated that they had never answered a question electronically in lectures, with a further 9.3% who had only answered the questions once or twice. Of these students who had never or seldom answered the questions electronically, 3.8% stated that they found the system useful even though they did not use it, leaving a remaining 12% who neither used it nor found it useful. Thus the BYOD strategy had proved highly successful as the technology and students' familiarisation with it matured over time. For all but a small proportion of students the interactive classroom system operating from students' own devices was recognised by them as a worthwhile innovation in the large lectures.

With respect to the equity concerns that are often raised with BYOD approaches, these were addressed in two ways, one a technological approach, the other a pedagogical one. First, the system was designed to minimise the cost to students of their interactions through a very plain student web interface which reduced data transmission. If students were unable, for whatever reason, to access the university wireless network to answer the questions, and had to answer using their own mobile service provider, the minimalist design meant that costs to them were very low. The calculation of transmission costs in 2008 showed that answering a single multiple-choice question using one's mobile service provider cost two cents when measured on a prepaid phone. With the lecturer normally asking three or four questions per lecture, this resulted in a cost to students of less than one dollar per semester. With changes to billing practices since 2008, costs have in fact declined and the cost of interactions is now regarded as a negligible part of most students' data plans: when an attempt was made to ascertain the cost of an interaction in 2013, it was so low that it could not be measured. Second, students have always been allowed to choose whether they answered individually or discussed the question with those sitting next to them and input one answer for the group. This is possible as students are not required to register to use the system and their answers are not graded, unlike the practice sometimes adopted with clickers. As well as removing the equity issue and allowing every student to participate, even if they do not own a device with Internet connectivity or are unsure about incurring transmission costs, this strategy fosters collaborative learning for those who choose it.

Therefore, the adoption of a BYOD approach to improving interactivity and feedback in large lectures proved successful and continues to the present day in this introductory accounting subject. It has also proved sustainable on a number of levels:

- It is economically sustainable, in that the university did not need to purchase clickers and the management software that is required to support them. Importantly, the university does not need to find the necessary recurrent funding for accessing the services of clicker vendors. Although there was a cost in developing the Internet-based interactive classroom system, since no software of equivalent technical sophistication was available at the time, this was a once-off development that fitted within university grant funding models.
- It is technologically sustainable, since students provide their own devices which they know how to use, without requiring support from the university, and which they will replace if their devices become obsolete or dysfunctional. The interactive system has proved robust and, should it fail, there are now very similar interactive systems available for free on the market which could replace the in-house system if necessary (e.g. *mQlicker*).
- It is pedagogically sustainable, since the lecturer was able to introduce an improved learning and teaching approach on a permanent basis without the logistic difficulties of distributing and collecting clickers and worrying about loss of devices in the process. From the viewpoint of the students, they have benefited from a more engaging style of lecturing, and have voiced their appreciation in surveys of receiving immediate feedback and correction of misunderstandings

Case study 2: Student-generated vodcasts as a means of providing multimedia content

In recent years there has been an increased demand for the provision of self-study learning resources to students. Certainly, major drivers of this demand include both the increased interest in MOOCs (Massive Open Online Courses) and flipped learning, where students study lecture content before the lecture and then the lecture time is spent discussing the pre-studied content. This is over and above the need for learning resources for distance education and e-learning that has been in existence since the introduction of the Internet to education in the 1990s. Even in face-to-face courses, supplementary materials are often provided to enhance opportunities for learning. This case study deals with the latter: the provision in a face-to-face course of non-standard learning resources which were found difficult to locate either in books or in a suitable form on the Internet.

The IT Careers Vodcast Project, which is the subject of this case, began with the recognition that students choosing to pursue a degree in IT often know very little about the types of employment to which their course of study is leading them. In order to address this situation, a careers assignment was introduced as a collaboration between the faculty and the university Careers Service in which students interviewed IT professionals in their workplace to find out more about their career. It was placed in a core subject, in which enrolment numbers are large, approximately 400 students in the first semester of the year when most school leavers enter university, and about 130 students in the smaller mid-year intake. Thus, like the accounting course discussed above, it was challenging in terms of the scale of enrolments.

In the first iteration of the assignment an attempt at sustainability was made by forming students into large teams of nine members each. In addition, only one student from each team went to the workplace to interview an IT professional about their career, with several students from different teams interviewing the same person simultaneously. This greatly reduced the number of IT professionals whom the Careers Service had to locate who were willing to be interviewed. Furthermore, the large team size decreased the number of presentations that the IT lecturer needed to schedule when teams were asked to present their findings orally to all the other students enrolled in the subject in a joint presentation schedule.

Following this first attempt, the assignment was withdrawn as it was deemed a failure from a pedagogical point of view. Ironically, the very measures that had been introduced in an attempt to make the project sustainable undermined its educational effectiveness (Dyson, 2014). For the students, the teams were too large so that many complaints about "loafing" were voiced since most of the work had been carried out by the students who conducted the interviews. Furthermore, these students were the only ones to gain the benefit of talking to an IT professional and seeing how the IT workplace functioned. Having these students *tell* their team members what they had found out was not nearly as powerful as the fieldwork experience itself. For the lecturer, the main issue was the logistics of organising a mass presentation in which all teams presented their findings to all other teams. Trying to find times that suited all students, given their different timetables, and a lecture theatre that was available was not easy.

The basic premise of the assignment, however, was still seen to be valuable and therefore was revived the following year in mid-2009, once the lecturer and tutors had had time to reflect on what had been learnt. Various pedagogical improvements were introduced including reduction of team size and allowing all students to take part in the fieldwork. The main change in terms of sustainability was the video-recording of the interviews in the workplace. The videos were shown in the students' tutorials and edited to create short vodcasts for sharing via the LMS with all students enrolled in the subject. This eliminated the need to organise a single large face-to-face presentation. A small university grant supported the reintroduction of the assignment, and this allowed for the purchase of some high quality, professional-grade video and sound recording kits. In addition, it facilitated the hire of a video expert, who located free resources on the Internet regarding appropriate file formats and editing. Links to these were made available to students, and continue to be used to the present day. Over time, the sharing extends beyond the semester in which students are enrolled as a library of vodcast resources about a wide range of IT careers has gradually evolved for viewing by students in subsequent semesters.

With the success of the IT Careers Vodcast Project, some further measures were introduced to ensure sustainability and scalability with the larger intake of students at the beginning of the year. The two challenges were locating sufficient interviewees, and providing enough video kits for all student teams. The first issue was solved through a request by the university Careers Service to the Australian Computer Society (the leading IT professional society in Australia) to assist with locating willing interviewees, in addition to using the faculty's existing IT industry contacts. Also, students were asked to find their own interviewees if they knew someone working in the field, while some part-time students who were already working in IT volunteered to be interviewed by other student teams. The second challenge regarding video equipment was solved by purchasing second-hand video cameras that another faculty no longer needed; the success of the assignment meant that funding was readily obtained for this. Moreover, the BYOD strategy was brought into play: some students used video cameras they or their families owned, or used their smartphone. The increased ownership of smartphones, as described earlier, meant that most teams had access to one and the video quality was quite sufficient for the purpose. (To reduce file size for upload to the LMS and easy download for viewing, students were advised against high definition video.)

The assignment has now been offered continuously, twice per year, since it was introduced as a student-generated vodcast project in 2009. It has achieved sustainability in the following ways:

- It is economically sustainable, in that students make the content for free for their peers to view, and enjoy being asked to do this. In addition, some students have been happy to follow the BYOD approach, even if many like using the high quality video recording kits provided, usually their first experience of using professional grade video equipment. The cost of purchasing new and second-hand equipment has been able to be borne by the faculty, given efficient arrangements for its hiring and return to minimise the quantity of equipment required.
- It is technologically sustainable, since students produce their vodcasts in standard file formats viewable on any computer.
- It is pedagogically sustainable, since the activity is highly motivating, provides rich learning experiences for the students and has resulted in statistically significant achievement of the target learning outcomes, as self-assessed by the students (Dyson, 2014; Litchfield et al., 2010).

For the lecturer, the main effort was in carefully designing the assignment and providing a comprehensive description for students to follow, and in hiring someone to source free support materials on the Internet during the first semester it was offered. The marking of the vodcasts is mostly an enjoyable and fairly quick task for tutors. Moreover, students work in teams, supporting each other on a largely self-directed project, with little need of help, apart from assistance from the Careers Service in locating interviewees at the beginning of each semester, which it sees as part of its role.

Conclusion

Though there are many challenges to achieving sustainability in mobile learning, there are also solutions to overcoming these challenges, if teachers and institutions are willing to embrace them. Two strategies for achieving sustainability, which are particularly suitable to higher education, are the BYOD approach to reducing or eliminating the cost of mobile devices, and adopting student-generated multimedia content as a way of providing resources to support students' learning. The application of these approaches to other areas of education is also a possibility. Certainly, student-generated multimedia, such as screencasts and slowmations, have been implemented successfully at the primary school level (Kervin, 2007; Rocha & Coutinho, 2011). However, BYOD is still not a reality in many schools due to policies banning student-owned mobile devices, particularly mobile phones, to manage what are perceived as ethical risks. However, these bans do not serve student interests well, either in terms of providing students with the benefits of mobile learning or preparing students for living in a mobile world (Dyson et al., 2013).

Both the BYOD and student-generated multimedia approaches fit well with widely accepted learner-centred pedagogies. They acknowledge students' agency in choosing which digital devices they should use in their education as much as in their private lives. Moreover, they recognise students' right to individual expression through multimedia content production, a practice consistent with the user-generated content that is now part of their world. The two case studies presented here demonstrate the sustainability of these two approaches economically, technologically and pedagogically. Sustainability has been achieved over a period of over five years and hopefully will inspire other teachers to adopt these innovative techniques

References

Adams, H. R. (2012). Bring your own device (BYOD) and equitable access to technology. *School Library Monthly*, *28*(8), 25–26.

AIMIA (2013). *Australian mobile lifestyle index report.* 9th Edition. AIMIA Mobile Industry Group, Australia.

Carnaghan, C., Edmonds, T. P., Lechner, T. A., & Olds, P. R. (2011). Using student response systems in the accounting classroom: Strengths, strategies and limitations. *Journal of Accounting Education*, *29*, 265–283.

Draper, S. W., & Brown, M. I. (2004). Increasing interactivity in lectures using an electronic voting system. *Journal of Computer Assisted Learning*, *20*, 81–94.

Dyson, L. E. (2012). Student-generated mobile learning: A shift in the educational paradigm for the 21st century. *anzMLearn Transactions in Mobile Learning*, *1*, 15–19.

Dyson, L. E. (2014). A vodcast project in the workplace: Understanding students' learning processes outside the classroom. In *Proceedings of the 13*th *World Conference on Mobile and Contextual Learning* (mLearn), Istanbul 2014, 258–271.

Dyson, L. E., Andrews, T., Smyth, R., & Wallace, R. (2013). Towards a holistic framework for ethical mobile learning. In Z. Berg & L. Muilenberg (Eds.) *The Routledge handbook of mobile learning* (pp. 405–416). New York and London: Routledge.

Dyson, L. E., Litchfield, A., Raban, R., & Tyler, J. (2009). Interactive classroom mLearning and the experiential transactions between students and lecturer. In *Same places, different spaces. Proceedings Ascilite*, Auckland 2009, 233–242.

Dyson, L. E., Raban, R., Litchfield, A., & Lawrence, E. (2009). Addressing the cost barriers to mobile learning in higher education. *International Journal of Mobile Learning and Organization*, *3*(4), 381–398.

Freeman, M., Bell, A., Comerton-Forde, C., Pickering, J., & Blayney, P. (2007). Factors affecting educational innovation within class electronic response systems. *Australasian Journal of Educational Technology*, *23*(2), 149–170.

Frydenberg, M. (2006). Principles and pedagogy: The two p's of podcasting in the information technology classroom. In *Proceedings of ISECON*, Champagne, *23*, 1–10.

German, R. F. (2013). The wild card character of "Bring Your Own": A panel discussion. *Educause Review*, Mar.-Apr., 11–28.

Habel, C. (2011). *VotaPedia* for student engagement in academic integrity education. *Ergo: Journal of the Education Research Group of Adelaide*, *2*(1), 15–25.

Herrington, A., & Herrington, J. (2007). Authentic mobile learning in higher education. In *Proceedings of the Australian Association for Research in Education (AARE) 2007 Conference*, Fremantle, 1–9.

Hoban, G., & Nielsen, W. (2013). Learning science through creating a "Slowmation": A case study of preservice primary teachers. *International Journal of Science Education*, *35*(1), 119–146.

Hoban, G., & Nielsen, W. (2014). Creating a narrated stop-motion animation to explain science: The affordances of "Slowmation" for generating discussion. *Teaching and Teacher Education*, *42*, 68–78.

Hoban, G., Loughran, J., & Nielsen, W. (2011). Slowmation: Preservice elementary teachers representing science knowledge through creating multimodal digital animations. *Journal of Research in Science Teaching*, *48*(9), 1–25.

Judson, E., & Sawada, D. (2002). Learning from past and present: Electronic response systems in college lecture halls. *Journal of Computers in Mathematics and Science Teaching*, *21*(2), 167–181.

Kennedy, G. E., Judd, T. S., Churchward, A., Gray, K., & Krause, K.-L. (2008). First year students' experiences with technology: Are they really digital natives? *Australasian Journal of Educational Technology*, *24*(1), 108–122.

Kervin, K. (2007). Exploring the use of slow motion animation (slowmation) as a teaching strategy to developing Year 4 students' understandings of equivalent fractions. *Contemporary Issues in Technology and Teacher Education*, *7*(2), 100–106.

Lambert, J. (2002). *Digital storytelling: Capturing lives, creating community*. Berkeley, CA: Digital Diner Press.

Laxman, K. (2011). A study on the adoption of clickers in higher education. *Australasian Journal of Educational Technology*, *27*(8), 1291–1303.

Lee, M.J., McLoughlin, C., and Chan, A. (2008). Talk the talk: Learner-generated podcasts as catalysts for knowledge creation. *British Journal of Educational Technology*, *39*(3), 501–521.

Lindquist, D., Denning, T., Kelly, M., Malani, R., Griswold, W. G., & Simon, B. (2007). Exploring the potential of mobile phones for active learning in the classroom. *Proceedings of the SIGCSE Technical Symposium on Computer Science Education*, March 7–10, Covington, KY, *39*(1), 384–388.

Litchfield, A., Dyson, L. E., Wright, M., Pradhan, S., & Courtille, B. (2010). Student-produced vodcasts as active metacognitive learning. In Proceedings, *10th IEEE International Conference on Advanced Learning Technologies* (ICALT), Sousse, Tunisia, 560–564.

McGarr, O. (2009). A review of podcasting in higher educaiton: Its influence on the traditional lecture. *Australasian Journal of Educational Technology*, *25*(3), 309–321.

Mohorovičić, S. (2012). Creation and use of screencasts in higher education. In *MIPRO*, Opatija, Croatia, 1293–1298.

mQlicker (n.d.). Retrieved on December 5, 2014 from: www.mqlicker.com

Mundy, D., Stephens, D., & Dykes, K. (2010). Facilitating low cost interaction in the classroom through standard mobile devices. In *Proceedings of the World Conference on Educational Multimedia, Hypermedia and Telecommunications 2010* (pp. 1819–1825). Chesapeake, VA: AACE.

Ng, W., & Nicholas, H. (2013). A framework for sustainable mobile learning in schools. *British Journal of Educational Technology*, *44*(5), 695–715.

Norris, C. A., & Soloway, E. (2011). Learning and schooling in the Age of Mobilism. *Educational Technology*, Nov.-Dec., 3–10.

Pachler, N., Bachmair, B., & Cook, J. (2010). *Mobile learning: Structures, agency, practices*. New York: Springer.

Patten, K. P., & Harris, M. A. (2013). The need to address mobile device security in the higher education IT curriculum. *Journal of Information Systems Education*, *24*(1), 41–52.

Prensky, M. (2004). What can you learn from a cell phone? Almost anything!" Retrieved on June 27, 2014, from http://thinkingmachine.pbworks.com/f/Prensky-What_Can_You_Learn_From_a_Cell_Phone-FINAL.pdf

Rocha, A. M. M., & Coutinho, C. P. (2011). Web 2.0 tools in high school in Portugal: Creating screencasts and vodcasts for learning. *US-China Education Review A*, *1*: 54–62.

Rossiter, D. E., & Crock, M. 2006. Embedding e-learning: A new perspective on change and innovation. *International Journal of Learning Technology*, *2*(4), 279–293.

Rubner, G. (2012). mbclick – An electronic voting system that returns individual feedback. In *Proceedings of the 7th IEEE International Conference on Wireless, Mobile and Ubiquitous Technology in Education*, Takamatsu, Japan, 221–222.

Scornavacca, E., Huff, S., & Marshall, S. (2009). Understanding the value of interactive SMS for large classes. In H. Ryu & D. Parsons (Eds.), *Innovative mobile learning: Techniques and technologies* (pp. 48–59). IGI Global: Hershey, PA.

Shafer, K. G. (2010). The proof is in the screencast. *Contemporary Issues in Technology and Mathematics Teacher Education*, *10*(4).

Tapscott, D. (1998). *Growing up digital: The rise of the net generation*. New York: McGraw-Hill.

Traxler, J. (2005). Institutional issues: Embedding and supporting. In A. Kukulska-Hulme & J. Traxler, (Eds.), *Mobile learning: A handbook for educators and trainers* (pp. 173–188). London & New York: Routledge.

Traxler, J., & Vosloo, S. (2014). Introduction: The prospects of mobile learning. *Prospects*, 44, 13–28.

Wheeler, S., Yeomans, P., & Wheeler, D. (2008). The good, the bad and the wiki: Evaluating student-generated content for collaborative learning. *British Journal of Educational Technology*, *39*(6), 987–995.

Index

For Product Safety Concerns and Information please contact our EU representative GPSR@taylorandfrancis.com
Taylor & Francis Verlag GmbH, Kaufingerstraße 24, 80331 München, Germany

www.ingramcontent.com/pod-product-compliance
Ingram Content Group UK Ltd.
Pitfield, Milton Keynes, MK11 3LW, UK
UKHW011000180926
475701UK00004B/484
9780815357490